FORESTRY, ENVIRONMENT & ECONOMIC DEVELOPMENT

Dr. (Mrs) Puspha Indurkar

ASHISH PUBLISHING HOUSE
8/81, PUNJABI BAGH, NEW DELHI-110 026

Published by :

S.B. Nangia
***for* Ashish Publishing House**
8/81, Punjabi Bagh
New Delhi-110026
Tele. 50 05 81
541 09 24

Show room :

5, Ansari Road,
Darya Ganj,
New Delhi - 110002
Tele. 3274050

ISBN 81-7024-443-9

1992

Typesetting by :
Computer Codes
Virat Bhawan, Comml. Complex,
Mukherjee Nagar, Delhi - 110 009
Phone : 7241200

Printed at :
Efficient Offset Printers
215, Shahzada Bagh Indl. Complex,
Phase-II, Phone : 533736, 533762
Delhi - 110035

PREFACE

In this book an attempt has been made to elaborate the theme of my Ph.D. thesis entitled "Contribution of Forest to the Economy of Nagpur District" on a microlevel. Efforts have also been made to include the data accordingly. Some of the chapters regarding Environment Protection, Social Forestry which concern indirect benefits have been elaborately discussed.

The depletion and degradation of the forest cover have caused great concern and everybody is after the solution for environmental degradation. The old National Forest Policy of 1952 has been unable to solve these problems. Therefore a new National Forest Policy of 1988 has been declared. This has been discussed in this book. Attempt has also been made to discuss the developmental works done in the Seventh Five Year Plan. Though the Eighth Plan has not yet been finalised, the draft proposals in the Eighth Five Year Plan have been given.

While writing and completing the information for this book, I was fortunate to get the required help from many persons. Foremost amongst them are Shri A.N. Ballal, Conservator of Forests (Evaluation), Nagpur and Shri Satish Elkunchwar with whom I had discussed on a number of occasions on various issues. I must express my unending gratitude to my husband Shri. R.N. Indurkar, IFS, Chief Conservator of Forests, Wildlife, M.S. Nagpur for his valuable suggestions and encouragement for this book. I must thank my sons Capt. Nithin, Pravin and my beloved daughter Ku. Mrunalini for their help, encouragement and forbearance, without which this work would not have been possible. Lastly I must thank Dr. (Mrs) Vijaya Katti, Lecturer, Indian Institute of Foreign Trade, New Delhi and Dr (Mrs) Kushalani, Lecturer, Dayanand Arya Kanya Mahavidyalaya.

I must thank Shri Pathratkar and Shri S.G. Patil for typing the manuscript and making it presentable.

I am quite confident that this book would be useful for the students studying environmental sceience and also for those who are concerned and interested in the forest conservation and protection of environment.

Dr (Mrs) Pushpa Indurkar

CONTENTS

LIST OF TABLES

LIST OF ANNEXURES

CHAPTER I

INTRODUCTION

1.1 Meaning of Economic Development

Development is a process of change in the present economic structure towards improvements. It is a function of conversion of the present situation towards betterment. It refers to Economic growth, Economic welfare, Economic progress and secular change. It deals with the problems of expansion of labour force, capital, volume of trade and consumption. It also involves the changes in techniques of production, social attitudes and institutions. Such changes may lead to economic growth. All these changes can be acquired through utilization and exploitation of the factors of production. Factors of production are the main Economic forces that influence development. Development process also needs proper direction, regulation and guidance to generate the forces of expansion and to maintain them. Natural resources are the main ingredients of the factors of productiion. For economic development the existence of natural resources in abundance is essential. In the words of Lewis, "other things being equal, men can make better use of rich resources than they can of poor". The problems of underdeveloped and developing countries like India, are closely related to exploitation and utilization of natural resources. The resources can be developed through improved technology and increase in knowledge. Lewis states : A country which is considered to be poor in resources today may be considered very rich in resources at some later time, not merely because unknown resources are discovered but equally new uses are discovered for the known resources. Nature has been very generous to India. She is endowed with variety of climatic conditions which enable her to develop variety of vegetation and crops. With different kinds of forests, she is also gifted with different kinds of wild life. Her mineral resources are varied and sufficiently rich. She is also fortunate to have a huge livestock. Thus, from the point of view of the availability of natural resources, India is one of the richest countries of the world. However, with all the available resources, India is still lagging behind in the economic development. There are large number of people below the poverty line. Unemployment is rampant. It is really a pity to find that there is poverty in this land of plenty.

1.2. Rural Sector in India

Till 1947, when we became independent, it was justified in attributing all the ills of the economic conditions to the exploitation by the foreign rule. But 44 years after independence and 40 years after launching our First Plan for development we cannot repeat the same argument. Planning for the development of India's economy started from the year 1951, when the first Five Year Plan started operating. In spite of the programmes undertaken under various five year plans during the past 40 years, the desired effects in the development of the economy are not visible. There are stll 48% people living below the poverty line. The growth in the economy has occurred in an inequitable manner. The gap between rich and poor has increased considerably. The lot of the rural poor is perhaps more pitiable than that of the urban dewellers because, the problems of the towns are concentrated and are constantly visible to the town-based administrators and politicians. As against this, the rural problems are diffused. The remedial measures require spreading of investment over a very large area. The severity of the problem facing Indian economy is further increased because more than 80% of the population in India is rural. Phenomenal rise in the population further aggravates the problem. It is, therefore, obvious that the real solution of achieving economic development in India lies in the development of rural economy.

1.3 Importance of Forestry in Rural Economy

The key for the uplift of rural economy in India lies in development of agriculture because 80% of the geographical area of the country is occupied by it and also nearly 80% of the total population is dependent directly or indirectly on agriculture. Forestry which is bracketed with agriculture has lot of potentialities for improvement of rural economy by providing innumerable goods and services and also the much needed employment. But forestry which occupies 22.7% of the total geographical area of the country is not given the due importance and attention it deserves. Most of the countries of the world (including India) regard forestry as part of the agriculture. First, because a forest provides both agricultural crops and wood and secondly, because forests play an important role in agriculture. But in the economy, forest and agricultural products play separate roles. Like agriculture forests also create new investment. Therefore, it is not fair to clamp the two sectors into one, and forestry should be considered as an independent sector. However, without going into the controversy whether forestry is an independent sector or part of agriculture sector it is clear that the role of the forestry as a vehicle for the uplift of rural economy is significant.

1.4 Rural Dependence on Forests

Forest trees provide fuel, small timber and other goods essential for meeting basic needs of the rural household. The forest and forest products generate income and employment in the rural community. Some of the benefits which the forest can bring to the rural community are summarised in Table 1.1.

Table 1.1

BENEFITS OF FORESTRY TO RURAL COMMUNITIES

Output	*Beneficial characteristics*
Fuel	(1) Low cost in use.
	(2) Producible locally at low cost.
	(3) Substitute for costly commercial fuel.
	(4) Substitute for agriculture residue.
	(5) Prevents destruction of productive resources.
	(6) Prevents diversion of household labour.
	(7) Maintains availability of cooked food.

Building Material	(1) Low cost in use. (2) Producible locally at low cost. (3) Substitues costly commercial material. (4) Maintains/improves housing standards.
Food Fodder Grazing	(1) Protection of crop land against wind and water erosion. (2) Complimentary source of food, fooder and forage (*e.g.* in dry periods) (3) Environment for supplementary food production (*e.g.* honey) (4) Increased productivity of marginal crop land.
Mailable Product	(1) Raising farmer/community income. (2) Diversify community economy. (3) Additional employment.
Raw Materials	(1) Inputs to local handicraft cottage and small scale industries (plus benefits from saleable product).

Source : NAO of AIO, Rome, 1978, Forestry for Local Community development

1.5 Role of Forests

(A) *Forestry and Employment*

An obvious element in most strategies for development is the choice of "Labour intensive" production processes and other form of economic activity. In a developing counting like India where the population of more than 800 million is continuously exploding and growing, the labour intensive techniques are more preferred. The forestry sector has ben recognised as a generator of employment because most of the forest operations are labour intensive. Seven million people enter the job market every year in India. Industry with all its downstream multiplier effect is not able to provide even

40,000 jobs annually. An exploding population compounds the problem. If the population grows at the current rate and jobs are not created any faster, the unemployed will rise to 166 million by 1994. The rural poor are without capital or land, the only thing they can sell is their capacity to do work. If employment is not provided to them, they are isolated from the distribution of wealth through the country's economic system. The Industrial Sector, both large scale and small scale, including cottage and mini Industries cannot significantly increase employment. Industry as it exists cannot solve the country's unemployment problem. It may even make the problem worse by diverting the country's resources to export without any benefit to the majority of the rural poor. Two hundred and fifty million people in rural India lack sufficient food and fuel because their earnings are inadequate. There are strong links between employment, poverty and a more equitable distribution of income, indicating the creation of employment is in itself an important factor for social and econnomic development. The only way to increase their income to meet their necessities is to provide employment in rural area in land use activities. There are hundred million acres of eroded, unproductive land all over India. This land can be converted into forest for which lakhs of rural people can get employment.

1.6 Rural Employment and Forestry

In development process more weightage should be given to rural India, so that the standard of living of the rural people may rise. This menas creating employment for the vast number of rural unemployed so that the earnings of a family of four are a minimum of Rs. 3000 per year or that every adult individual at present unemployed or underemployed earns at least Rs. 15000 per year. This is only possible in land use activities *i.e.* Agriculture forestry sector. Forestry activities should be synchronised in such a way that throughout the year people will get employment. Vast number of people can be absorbed in various forest operations such as Regeneration, Fire-protection, Plantation schemes and nurseries harvesting activities.

1.7 Environment and Ecology

With the rapid progress of Industrial development, human being have opened an era of rapid decline in environmental quality.

If unchecked, it will be impossible to maintain the earth's potential by the present generation for the future generations. The reduction in potential of the earth will go on increasing with the development process. But in the present decade situation has been changed and there seems a great awareness of environmental problems in all sectors of the society. People all over the world are deeply concerned with the preservations and protection of environment. In spite of the growing awareness and the strengthened administrative machinery, progress in environmental conservation is far from adequate—soil eroison and degradation, deforestation, siltation of rivers and lakes leading to frequent flash floods, loss of biological diversity and pollution of various kinds are all increasing. The growing investment in drought and flood relief is an index of the shape of things to come. It is, therefore, necessary to think of any type of development sustainable with the ecological balance.

1.8 Man and Ecosystem

Ecosystem is defined as the functional study unit of nature, since it includes both organism and abiotic environment. Environment is for the living systems of which human kind is an integral part. Nature and human beings form an inseparable part of the life support systems. The unlimited diversity of life forms on earth be the plants or animals including human kind is most remarkable. Modern human being first appeared several thousand years back. At first he was one among several species. But he could lead himself to the top of the heap because of his great intelligence, reason and scientific genius. Today mankind dominates the nature. He gives shape to the environment to suit himself and uses everything around him with scant regard whatsoever for the natural scheme of interdependent ecological regimes.

1.9 Balanced use of Natural Resources

In the development process we see the glittering material benefits showered by modern science and technology. This is like 'Alladin's lamp'. But this Alladin's lamp can turn out to be a pandora's box if not handed with care and wisdom. The chemical and other industries which are so vital for the health and well-being of modern man have also the potential of causing irretrievable damage to our greatest assets *viz.* land, air and water. The threat to the human race due to the environmental degradation is as great as the one from a nuclear holocaust.

Now there is a realisation that the future of mankind is intimately woven with ecology. Mankind has got to realise that future on this earth is bleak if he does not use natural resources rationally and does not take immediate steps to repair the damage already done to the ecosystem.

1.10 Forestry as an Ecosystem

Ecology is the science of interrelationship between organisms and their environment. The ecosystem is the basic functional unit of ecology, since it includes both environment and abiotic environment, each influencing the properties of the other and both necessary for the maintenance of life. Ecosystem is capable of self-maintenance except for the incoming solar energy and hence nutrient-species relationship is self-contained and cyclic. The socio-economic system of man is founded on a material base which is in contrast with the natural ecosystem. The increasing demands and aspirations of human society create a rising demand for goods and services. No regulatory function operates here in the form of market controls or natural laws. The perspectives of ecology are different from those of economics. The former stressess on limits rather than continuous growth, stability and continuous development.

Human being imposes changes on natural ecosystems and increasing control of his environment often creates conflicts between his goals and natural process. Forests which consist of flora and fauna play an important rate in maintaining ecological balance. But today human being has become blind in achieving the material benefits by destroying his own natural gift.

Today development has in a sense become synonymous with deforestation, desertification and progress with pollution. It is not imposible but very difficult to maintain the forest growth rate and increasing development rate at the same time and in the same area. Forest growth and development cannot go hand-in-hand. Establishment of big industries, big irrigation projects and agriculture development is being done at the cost of forest degradation . This does not mean that there should be no progress or development and that there should be no deterioration of natural resources like forests. Balance should be maintained between forest growth and economic prosperity (progress). Forest has two forest functions—(A) Long term ecological security, (B) Production of goods and services for the benefit of mankind. These two functions are called conservational/productional and productivity and exploitative forestry. The first forest function is known as, ecological role and the second one is called 'economic role'. The economic role of forest can be easily understood. But the ecological role is understood raw one with the shrinkage of forested area. The associate ecological degradation has become apparent.

Forest ecology studies reciprocal relationship of forest with environment. But considering two fold functions of forest, it has to meet both physical and economic needs of the population. So forest ecology combines in itself the study of all these factors that affect both natural and economic importance of forests. Since forests have the purpose of protection and production, forestry operations such as management, silvicultural treatments, forest hygienic operations, forest protection, selective felling, planting etc. introduce yet another factor, generally not fully appreciated in plant ecology and plant geography of natural tracts of vegetation. Forest ecology has thus to content itself very often with quasi natural conditions rather than natural.

1.11 Advantage of Managing Forest as Ecosystem

The ecological impact of forests is most easily recognised in their effect on micro climate and regulating the water regime of a catchment. Principles, techniques and methods of forest ecology find application in all those fields of human endeavour in which living plants are used as working material, whether singly or in groups. Thus applied forest ecology includes afforestation, soil conservation, desert and flood control, landuse, tank planning, river training, pasture, range management, nature conservancy etc. Forests in catchment area act as reservoirs by retaining water in the roots and layers and releasing it as a sustained flow over long period of time. Forests reduce soil erosion and thereby minimise the silting of the reservoir canals. The forests also provide habitat for wildlife. They also provide the recreational outlets for city dwellers.

Increasing realisation of the twin roles of forests, crash programmes wasteland development which include not only afforestation of the degraded forest, but also community forestry, reforestation, social forestry, agroforestry and energy plantation on non-agricultural land has been organised. Thus social forestry indicates the close relationship between social welfare and forestry. This new dimension of forestry for the upliftment of rural massess strengthens people-forest relationship not only to restore the ecological balance, but also advocates the society oriented forestry. Apart from the land under Government forest there are vast tracts of degraded lands which can be converted into forests for the proper maintaining of the eco-system. This work through the social forestry or community forestry activities can add to the generation of employment.

1.12 Awareness of Environmental Problems

Onwards 1960 there seems an awareness among local people, public in general and also in politicans. More and more funds are being allotted for the various schemes such as Social forestry, Wasteland development, Afforestation etc. Due to growing problem of pollution, environment problem is also being considered on government level and the severity of the problem is being recognised by all the sectors of Society. Due to industrialisation on large scale, the problem of pollution is becoming severe.

CHAPTER II

FORESTRY SECTOR IN INDIA

2.1 Area and Location

The total geographical area of India is just over 328.8 million Ha. The main land extends from 8° North to 37° North latitude and 68° East and 97° East longitude. It measures 3214 km. between extreme latitude and 2933 km. between extreme longitude. It is bounded in the South, West and East by sea with its vast line running to a length of 6100 km. It is the second most populous and seventh largest country in the world.

2.2 Climate

The whole Indian sub-continent comes under the influence of monsoon and therefore by and large is of monsoon type. The monsoon winds have widely influenced the climate of India. Though the climate is generally classified as hot in the larger part of the subcontinent, the changes in the climate from hot to temperate or even alpine is seen due to altitudinal variation in the Himalayan region.

As regards rainfall the regional distribution is uneven. On one hand there are heavy rainfall regions like Assam, Konkan and Sahyadris, on the other there are scanty rainfall areas like Rajasthan, Punjab. As we go in the interior away from the coast the rainfall decreases. Due to wide variation in temperature, rainfall and altitude; India is endowed with variety of climatic conditions.

2.3 Land Utilisation

As per recent available source the land utilisation pattern of India is given below :

Table 2.1

LAND UTILISATION PATTERN OF INDIA

Category	Area in million ha	% to total area
1. Agricultural and cultivable land	154.70	47.0
2. Forests with or without tree cover	75.18	22.8
3. Permanent pasture and grazing lands	12.15	3.7
4. Cultivation of tree crop and graves	3.91	1.3
5. Cultivable wastelands	16.64	5.1
6. Other marginal uses	17.53	5.3
7. Barren and Wasteland	24.60	7.5
8. Area for which no return exsits	24.09	7.3
	328.80	100.0

From the table, it will be seen that the major landuse is Agriculture. Forest landuse stands next.

2.4 Forests —Factors Influencing Distribution

In 1987, the recorded forest area in the country is 75.18 million ha. Climate is the most dominating factor affecting the distribution of natural vegetation. Dense forests are found in mountainous and inaccessible regions like Himalayas, Tarai, Assam and Sahyadris Satpudas, Vindhyas and eastern ghat areas. In other areas, there are continued attempts of illegal cleaning of forests for cultivation and settlement. In some areas, the illegal cutting of forests for domestic or commercial use have depleted the existing tree cover. With the result the actual tree cover is estimated to be 64.20 million ha. which works out to only 19.52% of the geographical area of the country. Out of this area only 36.14 million ha. is of adequate density. The effective forest cover is therefore limited to 10.88%. For ecological balance the country should have one-third of its land area under forests. It is observed that the existence of good forests is found where the rainfall is adequate and preserved and where the relief is unsuitable for cultivation. In the States like Madhya Pradesh, Assam and Orissa more than 35% of the land is under forests. However, in scanty rainfall regions like Punjab, Rajasthan it is less than 3%. Due to constant increase in the population more and more lands under forests were brought under cultivation. Apart from the pressure of the demand for agriculture, there are constant demands of the forest land for river valley projects, industrial estates, resettlement of project affected persons.

A country with an agriculture based economy, can thrive only, if it has a proportionate area under permanent vegetation. Forests affect soil climate and availability of water and thus sustain agriculture. Unfortunately, in India in proportion to land under cultivation the extent of forests is far from adequate. The comparative situation in some other countries is indicated below :

Table 2.2
COMPARATIVE SITUATION REGARDING LAND CULTIVATION OF VARIOUS COUNTRIES

Country	Total land Area million ha.	Agriculture land million ha.	% of Agriculture land to land area	Forests land Million ha	% of forests land to land area
India.	329	155	47.0	75	22.8
U.S.S.R.	2240	228	10.2	920	41.1
U.S.A.	313	186	20.0	290	31.8
Japan	37	4	10.8	25	67.6
U.K.	224	7	29.2	2	8.3
China	976	117	12.0	96	9.9
Thailand	51	16	31.4	21	41.2
Italy	29	9	31.0	6	20.7
Canada	922	44	4.8	326	35.4
Australia	762	45	5.9	107	14.0
Nepal	14	2	14.3	4	28.6

As regards the diversion of land from forest to other uses the National Commission on Agriculture has observed :

"As the administration of forests rests with the States, they are the ultimate authority to implement policy decisions. The provisions of policy decisions have not been fully implemented by them for various reasons. No systematic programme was drawn up to extend existing hire lands and establish new ones on lands belonging to Defence, Railways, Public works departments, University and Colleges, Local authorities etc., Relinquishment of forests land continued to occur for various purposes and the notion that forestry might be permitted on sufferance on residual land not required for any other purpose continued to hold swing."

2.5 Classification of Area under Forests

Table 2.3

CLASSIFICATION OF AREA UNDER FOREST

Description		Area in million ha. 1987
1. Legal Status		
(*a*) Reserved forest		40.18
(*b*) Protected forest		21.73
(*c*) Unclassed forest		13.27
	Total :	75.18
2. Actual Forest Cover		
(*a*) Dense forest		36.14
(*b*) Open forest		27.66
(*c*) Mangrove forest		0.40
	Total :	64.20
3. By Types		
(*a*) Tropical forest		52.00
(*b*) Sub-Tropical forest		5.77
(*c*) Temperate forest		4.57
(*d*) Alpine forest		1.86
	Total :	64.20

Source : Government of India, Ministry of Environment and Forest, Forest Survey of India, Dehra Dun. The State of Forest Report 1987.

Per Capita Forest area

With the population of 685 million in India, the per capital forest area comes to 0.1 ha. while per capita agricultural land is more than 0.2 ha. The per capita forest area compares poorly with the world average of 1.0 ha. The world average for per capita agricultural area is 0.3 ha. The following table indicates the Indian situation in comparison to other countries.

Table 2.4

PER CAPITA FOREST AREA IN HECTARE

Country	Forest land
India	0.1
U.S.S.R.	3.6
United States	1.3
Japan	0.2
United Kingdom	0.4
Thailand	0.5
Canada	14.2
Australia	7.6
Nepal	0.3

2.6 Types of Forests

On the basis of climate the forests of the country could be divided into four major groups.

I. Tropical forests

II. Montane Sub-tropical forests

III. Montane Temperate forests

IV. Alpine forests

2.6.1 Tropical Forests

(*i*) *Wet evergreen forests*. This is dense tall forest with height more than 45 metre or even more. It is entirely evergreen or nearly so.

This type of forests is mostly found in Arunachal Pradesh, Assam, Karnataka, Kerala, Manipur, Nagaland, Tamil Nadu, Andaman and Nicobar islands and Goa.

(*ii*) *Semi-evergreen Forests*. These forests mainly consist of deciduous species with certain number of predominant evergreen species. Therefore, these forests are less dense than the typical evergreen forests. The types are found in Assam, Gujarat, Karnataka, Andaman and Nicobar Islands and Goa.

(*iii*) *Moist Deciduous Forests*. In this type of forests the cover is mainly of deciduous species but the lower canopy largely consists of evergreen type of vegetation. The overall canopy is rarely dense and even. The average height is over 25 metre. This is found in Andhra Pradesh, Assam, Bihar, Gujarat, Karnataka, Manipur, Meghalaya, Mizoram, Tripura, Nagaland, Orissa, Tamil Nadu, U.P., West Bengal, Andaman-Nicobar islands, Goa and Dadara Nagar Haveli.

(*iv*) *Littoral and Swamp Forests*. It is mainly evergreen of varying density and height but always associated with wetness. It is found in Andhra Pradesh, Gujarat, Maharashtra, Orissa, Tamil Nadu, West Bengal and Andaman-Nicobar islands.

(*v*) *Dry Deciduous Forests*. It consists of entirely deciduous species or nearly so. The canopy is rather light with height 8 to 20 metres. It exists in Andhra Pradesh, Bihar, Gujarat, Haryana, Himachal Pradesh, Karnataka, M.P., Maharashtra, J. and K., Orissa, Punjab, Rajasthan, Tamil Nadu, Kerala, U.P. and West Bengal.

(*vi*) *Tropical Thorn Forests*. The forest cover is mainly deciduous with thorny tree species. The canopy is more or less broken with height under 10 metres. It is found in Andhra Pradesh, Gujarat, Haryana, M.P., Karnataka, Maharashtra, Rajasthan, Punjab, Tamil Nadu, and U.P.

(*vii*) *Dry Evergreen Forests*. These are mainly located in Andhra Pradesh and Tamil Nadu. The forests consist of hard leaved evergreens predominantly with some deciduous emergents. The forests are often dense with tree height less than 20 metres.

2.6.2 Montane Sub-Tropical Forests

(*i*) *Sub-Tropical Broad leaved Hill Forests*. In this, the forest cover consists of broad-leaved species which are largely evergreens. This type is found in Assam, Maharashtra, Meghalaya, West Bengal.

(*ii*) *Sub-Tropical Pine Forests*. These forests have certain number of pines in association. This type is found in Arunachal Pradesh, M.P., J. & K., Manipur, Meghalaya, Nagaland, Sikkim and U.P.

(*iii*) *Sub-Tropical Dry Evergreen Forests*. This is low xerophytic forest and scrub and is found in H.P., J. & K. and Mizoram.

2.6.3 Montane Temperate Forests

(*i*) *Montane Wet Temperate Forest*. This is evergreen forest without conifers. This is found in Arunachal Pradesh, Karnataka, Manipur, Nagaland, Sikkim and Tamil Nadu.

(*ii*) *Himalayan Moist Temperate Forest*. This is mainly evergreen with oaks and conifers. This is found in H.P., J. & K. and U.P.

(*iii*) *Himalayan Dry Temperate Forest*. This forest is open coniferous with sparse xerophytic under growth. This is found in J. & K. and Uttar Pradesh.

2.6.4 Alpine Forests

(*i*) *Sub-Alpine Forest*. This is stunted, deciduous or evergreen forest and usually in close formation with or without conifer. This is mainly in Arunachal Pradesh.

(*ii*) *Moist Alpine Scrub*. These are low in height but often dense scrub. This type of forest exists in H.P. and J. & K.

(*iii*) *Dry Alpine Scrub*. This is xerophytic scrub in open formation. This forest is mainly in Nagaland, Sikkim and U.P.

2.7 Statewise Forest Illustration of India

Forest Department statistics pointed out the 22,.7% *i.e.* 75 million hectares are classified as forest lands. Statewise distribution of forest land is given in table 2.5.

Table 2.5
STATEWISE FOREST AREA IN 1980

State/Union Territory	Forest area (thousand hectares)	Forest area as percentage of total area	Per capita forest area (hectares)
Andhra Pradesh	6,409	23.15	0.13
Assam	3,071	39.11	0.16
Bihar	2,923	16.81	0.04
Gujarat	1,952	9.96	0.10
Haryana	164	3.71	0.01
Himachal Pradesh	2,119	38.06	0.51
Jammu and Kashmir	2,189	9.85	0.37
Karnataka	3,787	19.75	0.11
Kerala	1,112	28.62	0.04
Madhya Pradesh	15,389	34.75	0.29
Maharashtra	6,408	20.82	0.11
Manipur	1,515	67.75	1.06
Meghalaya	855	38.02	0.66
Nagaland	288	17.42	0.41
Orissa	6,767	43.44	0.25
Punjab	243	4.83	0.02
Rajasthan	3,491	10.20	0.11
Sikkim	260	35.62	1.04
Tamil Nadu	2,179	16.75	0.05
Tripura	593	56.58	0.29
Uttar Pradesh	5,139	17.46	0.05
West Bengal	1,184	13.48	0.02
Andaman and Nicobar	714	86.13	3.90
Arunachal Pradesh	5,154	61.67	8.26
Chandigarh	Negligible	Negligible	Negligible
Dadra and Nagar Haveli	20	40.82	0.23
Delhi	Negligible	Negligible	Negligible
Goa, Daman and Dip	105	27.56	0.09
Mizoram	713	33.81	1.46
Pondicherry	Negligible	Negligible	Negligible
Lakshadweep	Negligible	Negligible	Negligible
Total	**74,743**	**22.73**	**0.11**

From Table 2.5, it seems that percentage of forest area to total area in different States is almost very low *i.e.* below 20%. Excluding in Andaman and Nicobar, Arunachal Pradesh, Tripura, Orrisa, Manipur, Dadra and Nagar Haveli which is above 40% and Mizoram, Sikkim, Meghalaya, Madhya Pradesh, Himachal Pradesh and Assam which is above 33%.

The national forest policy of 1952 aims at maintaining one-third of the total geographical area under forest in plains and 60% in hilly regions. But the real situation is different which is mentioned above.

2.7.1 Loss of forest land

Though forest area is mentioned as 22.73% of total geographical area, it is not a tree covered land. If means only that this vast area is under the control of forest department. There is no guarantee that it has tree cover. There is too much loss of tree cover from the legally constituted forest land. Prime Minister Indira Gandhi while addressing the Central Forestry Board in 1981, frankly admitted that only half of the official forest lands are under adequate tree cover. A Report of the National Committee on Environmental Planning also clearly states that no more than about 12% of the country's total land surface is under adequate tree cover. An areal survey conduced by the Indian Space Research Organisation in 1974 in Anantpur district found that many reserve forests in the district had no trees at all. "In addition a recent report prepared by the International Union for the conservation of nature and natural resources says that India has lost nearly 2.5 million hectares of mangrove forests over the last 80 years. These remarkable forests can grow on marginal lands and survive water-logging, soil infertility, salinity and heavy winds. But they have not been able to face the onslaught of human beings." All this is particularly depressing because it is in contrast to National forest policy of 1952 which advocated one-third land under forest. The loss of forest land is given in table 2.6.

Between 1951 and 1980, the total area that was brought under afforestation schemes amounted to a total of 3.18 million hectares of which 2.2 million hectares (69%) were covered by plantations for industrial purposes, only 0.39 million hectares (19%) for soil and moisture conservation purposes. According to forest department estimates of tree survival rates ranging from 50 to 70%. The total afforested area in the country during the three decades would not amount to more than 2.07 million hectares.

Table 2.6

FOREST AREA LOST BETWEEN 1951-1972

Purpose	Area (in thousand ha.)
River valley projects	401
Agriculture purposes	2,433
Roads and communications	55
Establishment of Industries	125
Total area	**3,402**

This represents an average arrived loss of 155,000 ha.

Source : Forest resources of tropical Asia, F.A.O 1981.

2.8 F.A.O. Report

A 1981 F.A.O report on forest resources of Tropical Asia clearly warns that the region is faced with a serious decrease in its forest stock. There is no evidence to predict that the great variety of forest services and functions, such as harbouring of wildlife, stabilisation of soil and of water and climatic regimes can be re-established in the foreseeable future. According to this report excessive deforestation leads to increasing floras, soil erosion, heavy siltation of dams built at an enormous expense and charges in and microclimate. In other words a progressive depletion of the country's ecological bank incessantly towards bankruptcy. Flood damages alone now average Rs. 1000 crores every year.

What is worse in human terms is the increasing denial to a large section of the country's population of forest products like firewood, which are vital for human survival. F.A.O. report warns that there is already an acute scarcity of firewood in the mountainous regions of northern India. Deficit situations exist in the indo-Gangetic plain and in Southern India. The remaining areas of India, which are not deficit in firewood at the moment are also expected to face a shortfall by 2000. A shortfall of 182 million cubic metres of firewood is predicted by the report by 2000 even if all planned plantations take place. Millions of women who have to collect this firewood daily are condemned to a grim future. The problem of overgrazing is also acute in India. The F.A.O. Report points out in 1975, India had about 15% of the world's cattle, 46% of its buffaloes, 17% of its goats and 4% of its sheep. "Grazing intensity is high in most of the forest areas of India. Incessant lopping of graziers has reduced vast forested areas to an unstable low scrub in the Himalayan zone. In addition to depletion of vegetative cover trampling by cattle is hardening the soil, preventing forest regeneration and causing soil erosion in several parts of the Indian Sub-continent." But the report also points out that the extensions of irrigated and cultivated land, areas is leading to a decrease in land available for grazing and in the absence of attractive employment opportunities, the graziers are being forced into a heavy reliance on grazing in forest areas and consequent degradation of the forests.

2.9 Causes of Degradation of Forests

The reasons for such degradation of forests are first a failure to meet the basic needs of rural people and second the exploitation of forests for urban markets. As people living near the forests are denied legitimate access to forests for satisfying their basic needs for fuelwood, building poles, fodder etc., they have had to resort to illegal and unorganised fellings or overgrazing. The maximum blame for deforestation rests squarely on the management policies of forest departments. "Forestry is not in essence about tree", says a leading forestry expert. It is about people. It is only about trees so far as they serve the needs of the people. But in India, forests have never been reserved for the people.

Many types of people have been blamed for forest destruction in India, including normals and graziers, tribal populations and shifting cultivators and ordinary village women who have to collect their daily fuelwood needs. The main sector of a country, *i.e.*, State Government which is responsible for the reduction in forest land should be blamed because Government always allots the forest land for agriculture to Adivasis on which agriculture crop cannot yield sufficient income to them, and the poverty among adivasis is increasing day by day because of their dependance on agriculture only and no work like forest operations, construction work etc. The configuration of the land, economic condition and the soil fertility are the main factors for deciding the utilization pattern of a particular land. Unfortunately this important aspect has not been considered by Government particularly in hilly area forest in Melghat of Amravati district of Maharashtra, where valuable teak forest was destroyed for agriculture crop. Terrain in Melghat is hilly which is very suitable for teak production. Almost entire Melghat is a hilly area, fertility is poor and land is shallow. It is a very difficult and expensive task to convert an infertile and shallow soil into fertile land. Colossal efforts and huge expenditure on crop on such a soil which is best suited for a natural regeneration of teak have been made. Is it a wise policy? If the answer is negative then unfortunately present development process will be considered a retrogative process. This has happened all over India. But since 1980 with the implementation of the Forest Conservation Act this situation has been finding its way from worst to better one, and not a single acre of forest land can be used or allotted to any institution or for any development process without the permission of the Government of India.

The other main reason for the depletion in forests is the short-sighted policies which were initially adopted by the British. Even after independence forestry sector is considered one of the main revenue earning sectors. The emphasis on commercial approach towards forestry cannot give justice to conservation-oriented outlook. Commercial outlook is the most dominant theme within the forest departments. In the year 1980-81, the total forest revenue of India was 464 crores from its forests with five States—Madhya Pradesh (Rs. 139 crores), Maharashtra (Rs. 52 crores), Karnataka (Rs. 46 crores), Kerala (Rs. 31 crores) and Orissa (Rs. 25 crores) according for 64% of this revenue.

Though forest officers follow the working plan systems which normally allow only the incremental growth to be felled without harming regeneration potential of the forest huge sections of forest are lost to illegal felling. Illegal fellings is rampant in various districts of the States.

Since independence it seems only an awareness toward urban needs and efforts are being made to meet the demand of industrial need. But the rural demand has been consistently overlooked at the same time. The expected production in 1990 of industrial wood will be about 14 million cubic metre and the current demand is about 24 million cubic metre. Forest department is consistently doing efforts to meet this demand. Naturally rural sector's economy which is forest oriented is being neglected and rural people are kept away from the forests. This type of attitude has been developed in policy makers. The proposed Forest Bill is an excellent example of such prevalling attitudes. As Madhav Gadgil, a leading botanist at the Indian Institute of Science in Bangalore, points out, the current economic situation is such that there is virtually no section of the country's population, the local population, the Government or the Industry which has personal stake left in the preservation of forests. "The local population is not bothered about preserving forests because it earns more if it steals the wood instead of working for low wages. The Industry is interested only in immediate profits. The forest bureaucracy has neither responsibility nor accountability."

2.10 The Present Condition of Depletion of Forests in Various States

2.10.1 Western Himalayas

Himalayan ranges are the best forest reserves of India. In the middle Himalayan belt which rises to an average height of 3000 metres, the forest area has been reduced to a mere 6 to 8% which was originally estimated at 1/3 of the total forest area. Below 2000 meters there are literally no forests left. The rate of depletion of forests in the Himalayan ranges is so great that this mighty mountain chain could become barren by the first half of the next century. According to a study of ecodevelopment in the Himalayan ranges conducted by the Planning Commission in March 1982, "Forest depletion is amongst the most urgent problems of the Himalayas that call for immediate remedial measures."

2.10.2 Jammu and Kashmir

The forest condition in this State is worst. The forest deteriorations set in after 1947. The forests were not regenerating, growing stocks had dwindled and some forests had been totally encroached upon and destroyed. Chirpine forests were totally destroyed. Commercial exploitation was the most powerful factor in the reassesment of the policy of cutting trees than the conservation approach. The extent and growth of commercial forestry in the

State can be assessed from the fact that whereas in 1947 the income from forest was Rs. 85 lakhs in 1980-81 it had grown to Rs. 34 crores. Due to domination of commercial purpose large size trees are seen rare. The other factor which is the main cause of depletion of forest is the construction of Roads without adequate soil conservation. Landslides are also caused by careless construction and the felling of large number of trees. Overall forests were damaged and have eroded the land so permanently that trees cannot grow there any longer. These roads are not useful throughout the year. They are only open for three months and for rest of the time are blocked either by snow or by landslides. Is this a sound economic policy of huge expenditure on roads and less on natural reserves for the gain of nothing, where the democracy and planned economy rules, and where the principle of "Maximum of Social benefit" is of utmost importance ? The answer is negative and the effect will be on the growth of economic development which will be pushed backward. While damaging the natural resource *i.e.* forest for the economic activity, "The principle of Marginal Productivity" must be considered. What is the benefit from the construction of roads through the forests which are blocked for nine or ten months and what is the ultimate cost of construction of roads in which the nature's valuable gift is damaged permanently ! The benefit is zero and the cost is infinity. The cost-benefit rate is not only in that the older trees were damaged but by cutting the matured trees as per working plan for the commercial purpose, the younger ones and growing trees were also damaged. Because of the heavy snowfall in J&K in the winter of 1979 large forest areas, specially the young trees were damaged. The older trees always protect the younger trees. So in such areas forest should be left untouched. In the words of Bahuguna, "When we disturb the forests and convert natural forests into monoculture timber fields, allowing the trees of only one age group to survive, they are victims of a calamity".

The excessive resin tapping which yields resin and turpentine is also the main cause for minimizing the chirpine forests. The trees are literally bled dry if the resin tapping is excessive. Consequently they weaken and break easily during storms. In 1981, wind broken trees provided 40,000 cubic metres of timber, which is higher than the prescribed annual yield of the J&K forests. "In J&K so extensive is the resin tapping that the State forests are likely to be exhausted within 10 years", says Bahuguna.

· Scarcity of a fuelwood because of the diversion of tree species for the forest based industries is the cause of the exploitation of other valuable species for the use of fuelwood. Demand for fuelwood is increasing rapidly. Srinagar alone consumes 46000 tonnes of fuelwood per year. Chief Conservator of Forests of Jammu and Kashmir predicts that soon the State will face a water famine because the hydel dams that exist in the State would then become "monsoon dams", near dry the rest of the year. Landslides, the result of deforestation are also covering up valuable crop land.

2.10.3 Himachal Pradesh

The same case discussed above is with this State. In mountainous region 60% land should be under forest cover. But in 1950 only 38.5% area was covered under forest in 1980. Study of the H.P. Government had concluded that the forest cover had fallen to a bare 18%. The Kulu valley, for instance, once a picture sea of Deodars-Dame 150 feet in height and 10 feet or more in girth is now almost bare. Nearly 1% of the land surface in H.P. has been stripped of its forest cover every year in the last 35 years. The forests have consistently fallen victim to the revenue hunger of this poor State. Rapid increase in revenue from Rs. 9.27 crores in 1977-78 to Rs. 16 crores in 1980-81 proves this fact. To assist the apple growing industry, the Government has given concessional timber rates to packing case manufacturers. They are allowed to set up saw mills inside forests and their usual practice is to clearfell the area. In this way 1.5 lakh cubic metres of timber are used each year, and by 1990 the Government estimates that this demand will rise to four lakh cubic metres and require the felling of 1.4 lakh trees every year. This is an enormous economic waste of valuable timber trees. Unfortunately soil erosion also takes place due to apple growing business.

Apple growing business is a profit making business. So many private land owners have cleared forested slopes to make way for apple orchards. As it takes five to seven years for an apple tree to bear fruit, farmers use interim to grow potatoes on the slopes of the forest land. For the immediate benefit of making revenue by the private landlords the valuable, indestructible, perishable forest wealth has been destroyed. Is it a sound economic justice in which the wealth has been concentrated in few hands at the huge cost of societies privileger ! If this wood were used for making furniture and quality paper, the value added would probably have been hundred times more. There never was a more urgent need to find a cheap and viable substitute for wood. Private contractor system is also responsible for the deforestation in all States. Though the trees are marked for felling, 5 to 10% unmarked trees are always cut by the contractors because the local forest guard can easily be bribed off with a bottle of country liquor. Besides this the tracts of forest is also closed for agriculture and tea plantations, large areas of existing forests are threatened because of indiscriminate resin extractions from chirpine trees. As Sunderlal Bahuguna puts it, "Broad leaved trees disappeared long ago from the Himalayan region, now even the chirpines are going down before the rapacity of a man." But in 1980 State was willing to put a complete ban on all commercial fellings for five years, but the Central Government was not agreed. Only contract felling was banned in nine out of the 27 forest divisions in 1980-1981 and according to the "Times of India" report the Government down also eliminated the remaining divisions by 1983. Government also made efforts for extensive replanting of coniferous forests. Upto 1984-85 the forest department made plantation rate to 70 million trees. They might just succeed however if they achieved this high rate afforestation in the current decade. After all afforestation, social forestry and wasteland development is not the ultimate remedy. There should be an emphasis on conservation policy rather than commercial policy. The destruction of the original forest done blindly cannot be overcome by planting the lakhs of new trees. Destruction of the forests damaged not only the trees but also threatened the whole environment. "Stop the cutting and preserve the nature's gift."

2.10.4 Uttar Pradesh

Forests in U.P. are now mainly concentrated in the eight northern mountainous districts collectively knows as Uttarkhand. The Uttarkhand region covers about 53,123 sq. km. Although about 37.45% of it is under forest over, about half of this forest area is degraded with poor tree density. The optimum density necessary for effective soil and water conservation is now found only in the remote valleys of the region. Nearly 8% of the total area is facing severe soil erosion and needs to be tackled on a priority basis. All over the region of Uttarkhand the old terror of tigers has been replaced by the even more severe terror of floods and landslides. The severe loss of lives experienced by the people in U.P. by the destruction of the forests led to a great movement of "Chipko Andolan". A new hydroelectric project and construction of roads have also become major causes of soil erosion. This State is also not free from resin tapping practices. Bahuguna states, "The condition of Rana Block in Yamuna forest division is most pitiable. We could see a large number of trees badly wounded with resin blazes on all sides. At some spots it seemed as if the area has been clear felled." The region has become a graveyard of pine trees.

The region of Uttarkhand also known as Devabhumi faces every year a major influx of pilgrims and tourists. This human pressure put heavy demand for fuel and hence pressure on forests. This pressure affected the valley of flowers by which the ecological balance rapidly changed. The number of flowers in this picturesque valley were going down every year. With the arrival of tourists and pilgrims in large numbers the unique flora of the valley is seriously threatened. Approximately 25000 tourists visited this region in 1981-82. The beautiful pastures of Turignath and Rudranath on the other hand, continue to be under heavy pressure from the local cattle population.

2.10.5 North-East Himalayas

This region is administratively divided into five States and two Union Territories, namely Assam, Meghalaya, Manipur, Nagaland, Tripura, Mizoram and Arunachal Pradesh. The North-east of India, is linked to the subcontinent by a narrow corridor sandwiched between Nepal and Bangladesh. This can be regarded as one entity because of its common features. The North-east represents one seventh of the total area under forest in India. It also plays host to a variety of birds and wild animals, including the famous one horned rhinoceros and to precious varieties of plant species. But the forest wealth of the North-east is beginning to disappear before the eyes of the people who have lived in its midst for thousands of years. Before the advent of roads, railway lines and so called economic development, the hills of the north-east and the fertile plains along the banks of the Brahmaputra, were covered with sub-tropical forests bearing large quantities of evergreens and deciduous trees. The people in the hills, over 150 different tribes, lived in harmony with the forests around them and used a form of agriculture that allowed them to cultivate the steep slopes. Now the forest areas are decreasing, human habitations are increasing and the fertility of the soil is diminishing at such a rate that the land is turning barren.

The total forest area is 8514 sq. km. This State was created out of three hill districts of Assam in January 1972. This is divided into three district councils *i.e.* Khari, Garo and Jaintia Hills which are elected with wide pioneers of governance under the sixth schedule of the Constitution. The forest area is 36.60% *i.e.* 8528 sq. km. in Meghalaya. This is below the 60% which is laid down for the hill areas by the National Forest Policy of 1952. The area under Reserve Forest is 706 sq. km. This is only 4% of the total area of the State. In effect, this is the only protected forest cover in Meghalaya. The remaining 7719 sq. km. of 90.51% of the State forests are unclassed State forests. They consist of forests directly managed by the district councils, community owned forests, village forests and forests owned by clans or individuals. This region abode of clouds has seen the thick, tropical forests covering its terrain reduced to shrubs. Now this forest cover is so degraded that today it is difficult to find a forest more than 50 years old.

Tribals resided in forest are granted tax exemptions and other benefits from forests, which are not available to non-tribals. A private ownership is permitted under the Schedule Sixty and no State authority can check what tribal owners of forests do with their trees. So most of the unclassed State forests are owned by tribals. They do the forest work through the non-tribal contractors ; they have to pay only 50% of the deposit required for fidding for a coupe of trees. They need show no Tribals are always bribed by the non-tribal contractors, who pay them a generous fee and operate that contract through them. For tribals this is easy money with no risk; for the non-tribals it is easy timber fetching huge dividends. But for the forests of Meghalaya's forest it has meant inevitable destructions. Today 800 licenced contractors converted this beautiful and picturesque hillock in a *bleak diad beauty forests,* which were once covered with Sal trees.

In Meghalaya land laws are not codified. The tribals have an oral traditions. People depend on customary laws handed down from generations to generations. So there is no record of land rights to indicate how much land is clan land, how much is community land or privately owned. Forests on such lands cannot be classified and therefore are open to unchecked exploitations. There is only one plywood factory, but there are innumerable saw mills, some operating within the forest area. Saw mills owners try to earn maximum profit and invest minimum ; district councils are not keen to take measures towards conservation. The reserved forests are managed according to a "Working Plan". The system of tree felling is scientific. This system has its own drawbacks but its advantage is that a vigilant and conscientious forester has the powers to check the extent of damage visited on a forest. The hilly terrain of Meghalaya coupled with the presence of numerous swift flowing rivers and the stable Geology makes it ideal for the development of hydroelectric power. In this hydel project which requires the construction of reservoirs for storing water, the excellent pine forests had to be submerged.

The system of jumming or shifting cultivation also leads to denudation of forests. The Jhumiya has reduced the cycle from 20 to 30 years to five to six years. So the forest cannot be regenerated in the fields left fallow for such a short time. The cycle of 20 to 30 years was the best for the growth of these hill region forests. The forests are victims of a vicious circle. Because, there are more people and less land for cultivations. Forest land that was once freely available has now been cleared for townships and other habitations. The impact of the various reasons seems in severe soil erosion and over time hills overlaid bare. Tops of hills are used to grow potatoes. It has resulted in an annual soil loss of 40 to 50 tonnes per hectare. These slopes become absolutely bare with no soil cover to the point that even grasses cannot grow.

Assam

The total forest area is 30,708 sq. km. According to official statstics 36.4% of the total area of Assam is covered by forest but in reality even as per forest officers the percentage is close to 19%. There are so many problems in Assam forests. The major problem is the politics of encroachment. Politicians have a readymade vote bank of people who are given legal rights to settle and cultivate the forest land. The forests of Assam became the cynosure of these politicians' eyes. They encouraged people to encroach into reserve forest. According to forest department figures between 1961 and 1981, 11,241 hectares of reserved forest were dereserved and given over for human settlement. In addition an estimated 1.36 lakh hectares of forest land are presently under encroachment. Besides immigrants, forests have been used to relocate flood affected people. But once a settlement is created inside a forest, it grows and eats into the forest area.

The other great problem in Assam is inter-State border dispute—Assam-Nagaland border and Assam-Arunachal border. The land in Assam is flat, whereas in Arunachal and Nagaland it is hiliy. Therefore, the tribals in Arunachal, Nagaland use this flat land for cultivation and now they have become encroachers. In this dispute hectares of forest lands are also destroyed.

The Government claims that there are 17.21 lakh hectares of reserved forests in Assam. But this does not account for the area encroached upon. The unclassed State forests, which are often added to the total forest area given as 30.71 lakh ha. are mostly inhabited areas with few trees and comprise 42% of the total forest area. Plywood lobby in Assam is a very dominant figure which is responsible for the denuadation of forest of the district Dibrugarh in Assam. Because of the area developing *i.e.* improvement in communications, roads and extension of the railway line, the establishment of plywood industry was proved feasible. The conditions were so favourable that plywood millers operating in West Bengal moved to upper Assam. In the year 1980-81 number of plywood factories were 52. They contributed to the State Government approximately Rs. 30 crores. Then cultivation in the two hills *i.e.* North Cachar hills and Karbi Anglong which have 40% of the State and total forest area is also contributing to deforestation. Oil and coal discoveries within reserved forests will also lead to clear the further forest tracts. Illegal felling of firewood is also there.

The consequences of deforestation in north-east area are severe. The main consequence is the extent of siltation in the river Brahmaputra and its tributaries. Deforestation in the catchment area has turned these tributaries into gushing torrents which lead the Brahmaputra with increasing quantities of silt each year.

In Assam the root cause of much of the problem has been the unnaturally high increase in population, because of the migration of people from Bangladesh. Unless people are given other sources of livelihood, they are bound to want at least a small plot of land: The pressure to occupy reserved forest lands, which are being used to supply timber to plywood factories is only understandable in these circumstances.

2.10.6 The Central Forest Belt

The Central forest belt comprises of Madhya Pradesh, Bihar and Orissa. This is the next largest concentration of forests after the Himalayan belt. These are mainly tropical most deciduous and semi-evergreen forests.

These forests are also the home of the majority of the 44 million adivasis of India who have drawn attention recently by launching movements on forest bases issues. Traditionally the existence of these tribals has been closely inter-forestry, it is their life that is most threatened by the ecological changes taking place.

2.10.6.1 *Madhya Pradesh*

The officially recorded area under forests in Madhya Pradesh is 15,389 thousand hectares (or 1,55,414 sq. km.) *i.e.* 35.1% of the total Geographical area. This State earns more thany any other State, Rs. 139.44 lakhs revenue per year. The Baster area which is rich in sal forest and largely inhabited by tribals was accepted for pilot project to increase the production of forests in 1976 with a five year 88.2 million World Bank funded. A good portion of the 22,000 sq. km. forest area was cleared rich sal trees and planted pine plantations. This project would have yielded 50,000 tonnes of pulp a year, sufficient to meet the needs of a new 250 crores paper mill. This was all done at the cost of Adivasis livelihood. For adivasis the sal tree is a part of their life. Sal wood is sturdier than teak. These tribals could earn money from the seeds of sal trees. The leaves of sal are used for making plates and bowls. This is all snatched by Hindustan Lever which uses the sal trees for making cocoa buffer used in chordate. Another local tree in Baster district *i.e.* Mahua is also a precious tree to the adivasis. The mono-culture would have brought development to the area, but only at the cost of valuable forests and adivasis. Once the natural forests are wiped out by commercial forests, it becomes a very costlier affair not only for the present generation but also for the future generation. This area is now reserved and based to adivasis. This area was proved unsuccessful for growing eucalyptus and pine. This soil is only good and rich for sal forests. Substitution of man made forests to nature's valuable gift is always a great damage to the human being. But human being always forgets this important thing though less efforts and huge expenditure is done for the short-sighted gains and loses all short and long term benefits, which is now experienced by every human being. Apart from this project illegal felling in the State is also caused to the damage of forests. This illegal felling was to the extent of 50 crores every year and honest forest officials were unable to control this privilege. Attacks on forest officials, rangers and guards by timber smugglers were so common that the M.P. Forest Guard Association threatened to go on strike unless full protection was provided by the State Government. Trains full of illegally cut teak continue to roll out of the forest regions of Raisen and Hoshangabad. The forest departments of M.P. and Maharashtra started a joint monitoring scheme to prevent this illegal trade. The M.P. Government also announced a ban on the export of firewood.

Former Chief Conservator of Forests, Waheed Khan pointed out that "Clear felling high quality sal to raise plantations of tropical pines is neither scientifically desirable nor industrially necessary".

2.10.6.2 *Bihar*

This State has a 16.8% land under forest cover. Total forest area is 29,227 km. This is mostly covered with Sal trees. But in 1977 replacement of the native sal by teak by the Forest Development Corporation created many problems for adivasis and also damage to rich sal forests. Because of the clear felling of this area Adivasis movement started to stop the clear felling. The movement was so bitter that even a forest official and forest guard could not enter the forest and the effect of this, the several other dense jungles virtually ruled by tribals. According to Times of India Report in February 1982, "Without anybody to check them, the tribals are indulging in indiscriminate felling of trees. Almost every second day there are reports that trees have been felled over another 50 or 100 acres". The current pattern of development as far as the Bihar tribals are concerned is such that it has snatched away their very environment from them, with which they co-existed for centuries but has given them nothing in return.

2.10.6.3 *Orissa*

The total forest area is 5.54 million hectares, *i.e.* 59,555 sq. km. and the well stocked forests covered only 2.83 million hectares, 22% of the total land area of the State. Forest destruction continues at a rapid pace in this State. Good forest cover is also affected by shifting cultivation, illegal felling and exploitation of forests for quick profits is a major problem in Orissa. Orissa State is considered as backward State. For the development of this State, Government sponsored projects have been started and these projects seriously threaten to disturb the Ecology of the unique forests situated in the picturesque Simlipal hills. These hills are covered with thick and impregnable rich sal forests. These forests also act as a massive water reservoir for the three contiguous districts of Balasore, Mayurbhanj and Keonjhar. Several major rivers originate from the Simlipal hills. In their untouched State, these forests still have a variety of fauna a few of which are also on the endangered list. The Man and Biosphere Committee of the Department of Science and Technology has suggested that Simplipal hills be conserved as a Biosphere reserve. Thousands of workers are engaged in collection of Sal seeds and the impact of human intrusion into the forest and the clearing of its natural undergrowth, can be ecologically disastrous. Just the fuel needs of these large numbers of workers will lead to a massive destruction of trees. This problem has not been considered by this State Forest Department.

2.10.7 Western and Eastern Ghats, which include Karnataka, Tamil Nadu and Kerala

2.10.7.1 *Karnataka*

Karnataka forests is enriched by valuable Bamboo forests. Total forest area is 38,645 sq. km. The development of paper mills is playing havoc with the Bamboo forests. Bamboo today forms about 67% of the raw material used by paper mills. A study conduced by Madhav Gadgil and S. Narendra Prasad for the Karnataka State Council for Science and Technology concludes that a large number of areas in Coondapur division and Koppa division, Sagar division, Chamarajnagar division and Mangalore division which were once rich in Bamboo Stocks are not totally barren. The major cause of this is the over-extraction by the paper mills.

The removal of bamboo stocks is having major environmental repercussions. Bamboo is very good soil binder. The reduction in Bamboo stocks leads to increased soil erosion. Ecologically bamboo is a sun loving species which can grow profusely wherever the canopy of moist or dry deciduous forest is opened up. But with the depletion of bamboo forest, this ecological niche is being taken over rapidly by a weed, particularly in the moist tracts. This also affects the regeneration of bamboo and all other species. The mangrove forests along the West Coast and almost all the area under evergreen forests is being consistently reduced.

2.10.7.2 *Tamil Nadu*

The same case happened with this State also as discussed above. This State covers 22,379 sq. km. total forest area. Thousands of poor villagers have suffered greatly because of the over-exploitation of bamboo stocks by the paper mills. They depend on bamboo for food, house construction, manufacture of agricultural implements such as seed drills and for making mats and baskets, an important source of livelihood. Instead of supplying the bamboo to the poor people for their livelihood, Forest Deptartment is favouring the paper mill. According to the Bombay Natural History Society (BNHS) study reveals the fact that "for the past three years, all the bamboo coupes have been allotted to the Mill". This Government policy deems to be an anti-socialistic policy.

2.10.7.3 *Kerala*

The total forest area in Kerala is 11,218 sq. km. Between 1940 and 1970, about 3500 sq. km. of forest lands in Kerala were diverted to non-forestry purposes. Large forest areas were assigned to the landless under the arable land scheme. In many cases rocky patches totally unsuitable for cultivation were found after the clearance of trees. The area under evergreen forests is being consistently reduced. The last representative of climax evergreen forests in India, the silent valley of the Palghat division of Kerala was recently threatened by a hydroelectric project. A major public campaign was able to save it. The land which was covered by forest but now shifted for cultivation has deteriorated rapidly and soil erosion has become a major problem, because of the negligence of Government in providing assistance for land improvement. Landslides have also become common in many deforested parts of Kerala. Efforts are being made at large scale for raising plantations of teak and eucalyptus. By 1976, the total area under plantations was 1310 sq. km. or about 14% of the total forest area. But Eucalyptus plantations have been almost completely wiped out because of fungal disease. The terrain of Kerala appears to be unsuitable for Eucalyptus plantations and is leading to severe soil erosion.

2.11 Growing Stock and Production

It has been estimated that 48% of the actual forest cover has a growing stock of 2260 million Cu.M. The projected figures of growing stock for the balance forest area is 1936 million Cu.M. Therefore, the total estimated growing stock of wood in the country is at about 4196 million Cu.M.

The recorded production of wood (timber + firewood) in 'the country during the last 10 years has varied from 26 million Cu.M. to 32 million Cu.M. The average annual production of wood in the country works out at 30 million Cu.M. The unrecorded production in the form of dead/dried and annual fallen wood which is removed on head loads for domestic energy is estimated at 22 million Cu.M. Thus, the average annual production within the silviculturally permissible limit is 52 million Cu.M. The net annual increment of 52 million Cu.M. is 1.24%.

The average annual production of wood per ha. is 69 Cu.M. A comparative study of average annual production of India with some other countries of the world is given below.

Country	Average annual production (Per ha. in M^3)
1. U.S.S.R.	1.9
2. U.S.A.	3
3. Japan	2.5
4. U.K.	3.5
5. India	0.7

The growing stock per hectare of the actual forest cover in the country works out at about 65 Cu.M. The World average is 110 Cu.M. per ha. In brief, in terms of both growing stock and annual production, the country is far behind the developed countries of the world.

CHAPTER III

FOREST POLICY 1988 AND FORESTY SECTOR IN SEVENTH AND EIGHTH FIVE YEAR PLANS

3.1 Forest Policy 1988

Forest is considered as the mother of a country. It is our treasure of genetic-pools. It is like a factory continuously producing a soil, oxygen, water, food and shade and other unnumerable services. This valuable wealth has been preserved by our ancestors and it is handed over to our present generation. But in the 20th century, forests have suffered a serious depletion because of pressures on forest land for non-forest uses and also to meet the demand of fuelwood, timber, fodder etc. National Forest Policy 1952, was inadequate to reduce the forest depletion. Forest being a state subject, there was no serious effort made for the preservation and conservation of forests. Commercial outlook always dominated for meeting the industrial demand without ensuring natural regeneration and compensatory afforestation. Politicians also used the forest land as one of the means to further their political interest. Maximization of the short-term benefits of economic development with the concept of sustainability, protection and conservation of forest land was almost forgotten. It has therefore become imperative to review and revise the 1952 forest policy with a view to evolve for the future a new strategy of forest conservation.

3.2 Objectives of Forest Policy 1988

The revised national forest policy called "National Forest

Policy 1988" was placed in the Parliament on 7-12-1988 and was approved in the House. The main objectives of this policy are :

(*a*) To ensure environmental stability and maintenance of ecological balance including atmospheric equilibrium which are vital for sustenance of all life forms, human, animal and plant.

This is the principal aim of forest policy. The derivation of direct economic benefit must be subordinated to this principal aim. This policy aims at 1/3rd of total land area of the country under forest or tree cover in plains and 2/3rd of the area in hills and mountains to prevent erosion and land degradation and to ensure the stability of the fragile and Ecosystem.

(*b*) To ensure the natural heritage of the country by preserving the remaining natural forests with the vast variety of flora and fauna, which represent the remarkable biological diversity and genetic resources of the country.

(*c*) Checking of soil erosion and denudation in catchment areas of rivers, lakes, reserviors, in the interest of soil and water for mitigating floods and droughts and for the retardation of siltation of reservoirs.

(*d*) Checking of the extension of sand dunes in the forest areas of Rajasthan and along the coastal tracts by stabilising the same.

(*e*) Increasing substantially the forest tree cover in the country through massive afforestation and social forestry programmes, especially on all denuded, degraded and unproductive lands.

(*f*) Meeting the requirements of fuelwood, fodder, minor forest produce and small timber of the rural tribal population.

(*g*) Increasing the productivity of forests to meet essential national needs.

(*h*) Encouraging efficient utilization of forest produce and maximizing substitution of wood.

(*i*) Creating a massive people's movement with the involvement of women, for achieving these objectives and to minimize pressure on existing forests.

(*j*) Minor forest produce should be protected, improved and their production enhanced to generate employment and income in the rural sector. In order to rehabilitate the degraded lands including forests emphasis has been given on massive afforestation scheme.

3.3 Strategy for Achievement of Objectives

Government has proposed the following strategy for achieving the national goal and above objectives.

3.3.1 Afforestation, Social Forestry and Farm Forestry

Attempts must be made on large scale to implement the time bound programme of afforestation scheme with tree planting of fuel and fodder development on all degraded and denuded lands in the country, whether forest or nonforest land, considering a national imperative.

Through social forestry programme planting of trees along the roads, railway lines, rivers and streams and canals and on other unutilized lands under state/corporate, institutional and private ownership programme of rousing in green belts in urban-industrial areas as well as in arid tracts must be implemented for checking soil erosion improving the micro-climate.

Considering the rural requirement for fodder, fuel and pasture, farm forestry programme should be developed on village and community lands including those of foreshores and environs of tanks not otherwise useful for productive uses. As the problem is enormous and efforts required would be colossal, the suceess of the programme would very much depend on the involvement of the local rural population. Government should provide the technical assistance and other inputs needed for this programme. The revenues generated through such programmes should belong to the panchayats, where the lands are vested in them. In remaining cases revenue should be shared by the local communities in order to provide an incentive. The ownership rights could be considered for the weaker section such as landless labourers, small and marginal farmers, scheduled castes, tribals, women etc. subject to appropriate regulations. Beneficiaries would be entitled to usufruct and would in turn be responsible for the security and maintenance of forests so created.

For intensifying this programme land laws should be modified wherever necessary. If possible degraded lands should be made available for this purpose either on lease or on the basis of tree patta scheme as per land grant rules and land ceiling laws. Felling of trees on private holding should be governed with appropriate regulations.

3.3.2 Management of State Forests

The State Government should follow the Central Government guidelines in giving permission for working on forest land and for giving the approval of the management plan for any forests. Any management plan should be in a prescribed systematic forest and should be as per national forest policy.

Tree clad land with steep slopes, catchments of rivers, lakes and reservoirs, geologically unstable terrain and such other ecologically sensitive and fragile areas must be preserved and interference of schemes and projects must be severely restricted. Tropical rain forests situated in Arunachal Pradesh, Kerala and Andaman should be totally safeguarded.

To reduce the gap between demand and supply of fuelwood and other needs, it is necessary to enhance the forest cover through the application of scientific and technical inputs and production forestry programme should be aimed to reduce this gap. But at the same time it is essential to follow long-term scientific trials undertaken by specialists in Ecology, clearfelling of adequate stock of natural forest and not to introducing exotic species through public and private sources. There should be no adverse impact on native vegetation and environment by introducing the programme of clear-felling or plantation.

3.3.3 Rights and Concessions

Forest capacity is always limited to unable to meet the growing demand of rights and concessions and also grazing by the cattle. So forest capacity should be optimized on the basis of the increased investment, silvicultural research and development of area. The gap between forest capacity and the growing demand should be met by encouraging social forestry programme outside the forest as well as introducing the stall feeding of cattles. Rights and concessions are strictly for the people who live in forest area particularly for tribal people as well as for the weaker sections of society. The right holders should be motivated to identify themselves with the protection and development of forests from which they derive benefits. Forest should be primarily for the bona fide use of communities living within and around forest areas. Their domestic requirements of fuelwood, fooder, minor forest produce and construction timber should be made available at reasonable price at conveniently located depots.

Because of the short supply of wood, there should be a long term solution to minimise the gap between wood supply and demand record. This problem can be solved by introducing the substitute for wood used in furniture, railway sleeper, construction of building and industry etc. The alternative source like biogas, L.P.G. and solar energy, fuel efficient 'chullas' should be used for the substitution of fuel needs. There is also a need of popularising the alternative sources through demonstration and other means.

3.3.4 Diversion of forest lands for nonforest purposes

Forest wealth is owned by entire community. So the entire tree cover, or forest land should not be made easily available or utilised for various projects and programmes. This national asset should be strictly safeguarded for providing sustained benefits to the whole nation. The nature has taken a considerable time to build up this resource and the same should not be used just for achieving political and local interests. The diversion of forest land to nonforest purposes like dam, projects, etc. should be subject to most careful examinations by specialists from the point of view of environmental cost and benefits. Compensatory regeneration and afforestation schemes should be sponsored and funds for these schemes should be made available by the industrialists and Government departments using forest lands for their projects.

Beneficiaries in mining and quarrying in the forest land and tree covered land should bear the responsibility of repairing and revetating the area in accordance with established forestry practices. Proper appraisal of mine management plan from the environmental angle in needed to grant lease mining to any party whether it is private or public.

3.3.5 Wildlife conservation

Forest is a genetic pool to maintain genetic continuity. Forest management should include special care of the needs of wildlife conservation. It is specially essential to provide for "corridors" linking areas protected for wildlife to ensure free movements.

3.3.6 Tribal people and forests

There is a close symbolic relationship between tribal people and forests. Considering their dependance on forest, they should be identified as part and parcel of forest. Therefore it is the duty of all agancies responsible for forest management including Forest Development Corporations to ensure their participation in the protection, regeneration and the development of forests. Degradation of forests by illegal cutting by the private contractors should be minimised by replacing the contractors by the tribal labour cooperative societies. These societies should carry out the forest activities under the guidance of forest officers as per rules and regulations laid down by the Government. In fact Tribal Labour Cooperative Societies should act as caretakers of the entire tribal community to provide minor forest produce and other facilities of essentials. Efforts should be made for the upliftment of tribal community through forestry oriented schemes with the help of tribal beneficiaries and integrated area development programme.

3.3.7 Shifting Cultivation. Damage to Forest from Encroachments, Fires and Grazing

In hilly region environment and productivity of land is adversely affected by shifting cultivation. So the area which is already affected by propagating improved agricultural practices should be allotted for such type of cultivation and affected land productivity and environment should be improved and maintained by rehabilitating the area through social forestry and energy plantations. Encroachments on forest land should not be regularised. Effective action is needed to prevent encroachment on forest land.

The incidence of forest fires is high in the country. To reduce the destruction of standing trees, fodder and natural regeneration by fire, improved and modern management practices should be adopted.

Grazing in forest areas is a severe problem. It should be solved with the active help of community involvement. Special conservation areas, young plantation and regeneration areas should be fully protected. Grazing of large herds of non-essential livestock should be prohibited by levying adequate grazing fees so that keeping such non-essential cattle should be uneconomical.

3.3.8 Forest Based Industries

The forest policy emphasises that the forest based industries should try to meet their demand of raw material as far as possible from the areas outside the forest by establishing direct relationship between the factory and the individuals who can grow raw material for them. The industry should in turn support the individuals with imputs like credit, technical advice and havesting and transport services. The arrangement is advocated mainly to see that the fuel, fodder and timber requirement of the local population is not sacrificed for providing raw material to industries.

Forest based industries should consider the interest of the local people by providing them employment and at the same time ensure their involvement in raising trees and plantations.

Marginal and degraded lands and not the natural forest land should be used by small and marginal farmers to plant wood species required for industries. These may also be grown along with fuel and fodder species on community lands not required for pasture purposes and by forest department/corporations on degraded forests not earmarked for natural regeneration. Use of alternative raw material should be encouraged. Supply of forest produce at concessional rates should be stopped. Import of wood and wood products should be liberalised. All these considerations will however be subject to current policy, relating to land ceiling and land laws.

3.3.9 Forest Extension

Realisation of the importance of forest conservation to individual, as well as the whole community is the need of the day. No programme can succeed without the full devotion and sacrifice of the community. Therefore, cooperation and willing support of the massess in this programme is essential. Community involvement should be inclined towards the mass forestry programme through training, lectures, and demonstrations of planting trees on a large scale. A part of forestry training should be included in the curiculum of primary schools to colleges. Farmers and interested people should be provided opportunities through institutions like Krishi Vigyan Kendras, trainers training centres to learn agrisilvicultural and silvicultural techniques to ensure optium use of their land and water resources. Short-term extension courses and lectures should be orgainsed in order to educate farmers. Suitable programmes should also be propagated through mass media, audio visual aids and extension machinery.

3.3.10 Forestry Education and Research

Forestry should be recognised both as scientific discipline as well as profession. Forestry education, research, training and post-graduate research, professional excellence should be facilitated in such a way that large number of qualified professionals will be available to implement the forestry programme with up-to-date modern techniques and knowledge. This training and knowledge should cover productivity, regeneration of barren, marginal and wasteland watershed areas, conservation and management of forest resources, tribal development research, development of substitutes to replace wood and wood products and research related to wildlife preservation, management of national parks and sanctuaries.

3.3.11 Personnel Management

In matters of personnel management the Government should aim at increasing the professional competence of profession, foresters and forest scientists. It also aims at attracting and retaining qualified and dedicated personnel who have to work in the remote and hospitable places.

3.3.12 Forest Survey of Data Base

Survey of Forest Resources on scientific lines and to up-dating information should be done on priority basis to provide reliable facts and to reduce false sense of complacency because of inadequate data regarding forest resources. Use of modern technology and equipment can improve the quality of periodical collection and publication of reliable data on relavant aspects of forest management.

3.3.13 Legal Support and Infrastructure Development

Effective implementation at the Centre and State level needs appropirate legislation supported by adequate infrastructure.

3.3.14 Financial Support for Forestry

Achievement and souccess of new forest policy depends upon the lot of investment of financial and other resouces on a substantial scale. Adequate funds should be made available for forest growth and development which is fully justified considering the contribution of forests in maintaining ecological processes and life support system and in preserving genetic diversity. Forest wealth is a national asset which should be protected and enhanced for the well being of the people and the Nation.

3.4 Appraisal of the Forest Policy 1988

No doubt this forest policy 1988 is a comprehensive document. This policy recognises the importance of forest conservation approach which aims at the preservation and protection of the national wealth which is handed over by the ancestors. It also considers the enhancement of forest land and tree cover by raising plantations, by implementation of various schemes such as massive afforestation programme, social forestry and farm forestry etc.

This policy emphasis the 1/3% forest land in plain area and 2/3% in hilly region of total geographical area which will help to maintain the ecological balance and preservation of environment. To achieve this object, it has been prescribed that there should be no such programme of reducing the tree cover by clearfelling and introducing new exotic species for narrowing the gap between demand and supply. This suggestion should be appreciated by the whole country because for the first time such type of strategy is introduced. It has already been expereinced that the introduction of exotic species like eucalyptus everywhere has created problem.

It is a welcome suggestion of emphasising the Afforestation, Social forestry and Farm forestry schemes on large scale which will be helpful in maintaining ecological balance and for satisfying the social need regarding fuel, fodder etc.

It is true that people living in forest and around the forest should have privilage of customary rights and concessions and these rights and concessions should be protected and preserved. But at the same time history reveals the fact that they are aware only in fulfilling their rights and not in their duty of protecting the forest. Duty and rights are the two sides of one and the same coin and whenever they enjoy their rights they should also safeguard the forest wealth for their own interest. So it is the duty of the government, local social and dedicated institutions, politicians to make them aware of their duty of protecting the forest.

There is always conflict between economic development process and growth of forest. To achieve the objectives of five year plans large chunk of forest land are diversified for various big projects, industries, dams, etc. The policy document suggested the compensatory afforestation programme and huge scheme of establishing green belts. But this man-made forests cannot be substituted for original natural forest. But if the principle of "maximum welfare of the maximum people" is to be achieved by establishing the big projects and industries, mining etc. then the priority will be towards these activities. But regarding forest wealth, there should be minimum sacrifice in the form of diversification of forest land. It is the general experience that in the beginning there is a lot of enthusiasm of planting trees but after some period maintenance is ignored. The species which are selected for green belt or afforestation programme should also be indigenous local species which were destroyed in making these projects. In big projects and dams lakhs of hectare of forest cover is destroyed. Compensative plantations schemes are not enough to regain ecological loss.

Nobody will deny the close relationship of tribals and forests. But since independance under the tribal development programme lakhs of rupees are invested, the valuable forest land is diverted to agriculture use for raising their level of income. But what is the result ? There is no such noticeable improvement in their condition. Today also they survived by the roots and fruits of plants. Forest labour cooperative societies have been established but these societies are dominated by the tribal leaders and the poor tribals are being exploited by these leaders. To develop the forest village on par with revenue village, the change in the orientation of planning process is needed. Undue importance should not be given to agriculture. The whole family is engaged on agriculture work on the holding, ranging from 2 acres to 16 acres. The farm on which they cultivate is uneconomic holding. The configuration of the land, economic condition and the soil's fertility are the main factors for deciding the cultivation pattern of a particular land. While developing the forest village it is essential to take into consideration the land capability pattern and prescribed utilization of land accordingly.

Individual beneficiaries scheme and area development programme are being implemented under five year plan. But only few privileged Adivasis and their relatives are getting the benefits and most of the Adivasis are still below the poverty line. Area development programmes like schools, dispensaries etc. are being implemented for the progress of tribal people. While doing so forest area is reduced and many a time the infrastructure so developed remains without utilization for want of proper personnel to run them. It is, therefore, very uneconomical as well as not practical to spend lakhs of rupees on such programme. Therefore, their development process needs an altogether different approach in which both objectives *i.e.* forest protection and development of tribal economy will be achieved. To make the area development programme economical as well as practical it is necessary to consolidate the smaller villages in big villages and then provide them with the basic amenities. With such approach it will be easier to provide the modern basic amenities under area development prog-ramme in a concentrated manner instead of dispersing them at several places. In addition to this, those lands which are not fit for agriculture should be used for plantations of forest species or horticulture.

Encroachment problem is a severe problem. The strategy of government regarding encroachment is proper. However this needs to be strictly implemented without any political interference.

Government has not considered the severity of grazing problem. It is the grazing which is the enemy of the forest. If government has the political will they can stop grazing of cattle in forest area and save the forest from further degradation. The government should encourage people for stall feeding their cattle and cutting grass from the forest area. The supply of grass from the forest area can be at very low cost or at heavily subsidised rate.

Domination of commercial approach led to the growth of forest based industries. Forest based industries consume lots of forest produce in the form of raw material at the cost of local people. This policy document emphasises on the forest conservation approach. It is advised in the forest policy that the forest industries should use raw material other than forest produce. The policy document has given more priority towards the meeting of local demand of forest produce. Forest produce being limited a proper balance has to be achieved to supply these industries and to supply for meeting of local demand.

This policy aims at forest extension programme with willing support and co-operation of people. It is true that only people's participation on a large scale will give success in regaining the lost forest wealth and in preserving the present forest land and tree cover. Therefore the suggestion of organising programme of short-term extension courses and lectures, demonstrations to educate farmers and use of mass media, audio-visual aids and extension machinery is a welcome suggestion.

The best management needs good adminstration. Good administration leads a good implementation of the policy which can help in maximising the forest growth. This is possible only through forestry education and providing research facilities etc. It is therefore necessary that all the existing Training Institutes particularly the Indian Institute of Forest Management at Bhopal need to be strengthened to shoulder new responsibility as indicated in the policy document.

The main forestry research institute is at Dehradun. If these researches made there are to be denominated for the use of the common man then it is necessary to have small centres at district level like agriculture.

The government recognises that the forest officers job is arduous and also expects full dedication from them. The dedication to the duties while working in the interior forest areas is possible only when the government shows some consideration to the diffculties faced by the foresters in problems of education of their children.

The recommendation regarding not treating forest as a source of revenue is most welcome as it emphasises the strategy of forest conservation.

3.5 Appraisal of Forestry Sector in The Seventh Plan

During the past thirty years, there has been considerable activity in the field of economic development. Large amount of money has been spent by Governments, international & bilateral agencies in the pursuit of economic growth. Forest development is one of the most important parts of Economic growth, but the experiences shows that process of Economic development leads to reduction in forests rather than their development for achieving immediate advantages.

3.5.1 Rural Problem

The roots of the problem of rural poverty are, in general, population growth and rising expectations. The spread of information has led country dwellers to increase their demands and to have the benefits that they now know, are enjoyed by many in the towns. In thickly populated regions, rural people have sacrificed their forests, since weed is less indispensable than food, though in the long run the absence of woodlands may reduce farm ouput. This has often led to soil erosion, where agricultural crops have been pushed on to unsuitable land. Temporary relief from feed shortages has thus been brought by consuming the biological capital of trees and soil, leaving a smaller capital for future production of all kinds. This resulted into land degradation and created a serious Ecological and Secio-Economic crisis in the country. This crisis has led to new approach to forest conservation.

3.5.2 Forest Conservation in the Seventh Plan

More emphasis on the conservation of forest started from the sixth five year plan. It has marked the beginning of integrated and intensive efforts in conserving our natural assets. These efforts have been given more importance in the seventh five year plan also. It laid even greater stress on an integrated approach to the conservation of land, water, flora and fauna in addition to the prevention of atmospheric and terrestrial pollution. To achieve the objectives laid down in the seventh five year plan regarding Forest and Environment, Wildlife and Ganga Project, the Ministry of Environment and Forests was created in the year 1985; the first year of the seventh plan. The allocation of outlays in the seventh plan under different sectors are as under :

Table 3.1

THE ALLOCATION OF OUTLAYS IN THE SEVENTH PLAN

(Rs. in crores)

	Central Sector	State Sector	Total
Forest and Wildlife	446.71	1412.30	1859.10
Environment	110.10	77.91	188.01
Ganga	240.00	-	240.00
Total	796.81	1490.30	2287.11

Source : Ministry of Environment and Forests.

3.5.3 Sectorwise Break-up of Financial Position

As per above outlay of central sector the sectorwise break-up of the financial position is an under :

Table 3.2

SECTORWISE BREAK-UP OF FINANCIAL POSITION

Sector	Seventh plan outlay	Expenditure during the first three years (1985-86 to 1987-88)	Outlay during the last two years (1988-89 to 1989-90)	
1	2	3	4	
			1988-89	1989-90
Environment	110.00	60.66	30.00	33.79
Ganga Action Plan	240.00	75.33	57.00	64.00
Forests and Wildlife	155.00	48.97	28.00	31.75
National Wastelands Development Board (NWDB)	292.00	124.26	66.10	72.55
Total	797.00	309.22	182.00	202.00

Source : Ministry of Environment and Forests.

3.5.4 Enviornment and the Seventh Plan

The Ministry of Environment and Forests have executed the programme of Environment, Ganga Action Plan, Forests and Wildlife and National Wasteland and development in the seventh plan. The principal thrust of all these programmes during the seventh plan is to ensure environmental stability which is so vital for the ecological balance and conservation of forest. These programmes are also being implemented by the States within the outlays provided.

3.5.5 Environment Programme

Environment sector includes Pollution, Monitoring and Control, Environmental Impact Assessment, Conservation and Survey, Eco-regeneration, Research, Development, Environmental Information, Environmental Policy and Law including Management and Control of Hazardous substances and International Co-operation Programmes.

3.5.6 Ganga Action Plan

The Ganga action plan includes schemes of low cost sanitation, river-front development and construction of electric crematoriums.

3.5.7 Environmental Stability

In ensure environmental stability and maintainance of Ecological balance including atmospheric equilibrium the forest policy 1988 has been fermulated by the Government of India.

3.5.8 Forest Conservation

Forest conservation has been promoted by vigorous implementation of the Forest (Conservation) Act 1980. Forest Conservation Amendment Act, 1986, introduces the schemes for protection of forest from biotic interference. A pilot project for controlling forest fires using new methodologies is being implemented in Utter Pradesh and Maharashtra in this Seventh Plan. This scheme appears to be successful and thorefore it has been proposed to extend this project to additional areas.

3.5.9 Forest Survey of India

The Forest Survey of India has been strengthened to prepare a national vegetation map bi-annually. A National Natural Resources Management System has been initiated.

3.5.10 Research Activities

To provide proper impetus and thrust to research activities and education, forestry research has been re-organised. Indian Council of Forestry Research and Education has been constituted. Five research institutes have been set up at five places to carry out research in particular facets of forestry and to serve the regional needs. Refresher training for forest officers has been introduced. The Indian Institute of Forest Management and the Wildlife Institute of India have also been established.

3.5.11 Wildlife Management

Modern Wildlife management requires comprehensive amendments. So previous Wildlife (Protection) Act 1972 is being examined for bringing improvement in new amendment. This comprehensive amendment may help to make more responsive implementation. 'Project Tiger' scheme has also been expanded by establishing more tiger reserves. The network of National parks and sanctuaries has been expanded with 421 National parks/Sanctuaries covering 1, 33.00 sq.km. and 4% of the nature's area. Projects were taken up for conservation of rhines and snow lopards (Report, Ministry of Environment & Forests, page. 6).

3.5.12 Afforestation Activities

In the Seventh Plan, to expand the afforestation activities on large scale through various schemes the National Wasteland Development Board was established in May 1985. 5.07 million hectares land have been planted in the last three years. Apart from planting, this board focuses on halting degradation and on improving the productivity of degraded lands. New strategy for preparation of village level based plans, greening of the Himalayas, demonstration of existing technologies for reclamation of Wastelands on a large scale is formulated and it is being implemented to achieve the aim.

3.15.13 Physical Achievement

The Physical achievement of the Seventh Plan regarding the progress under the 20-point programme is as follows.

Table 3.3

PHYSICAL ACHIEVEMENT OF THE SEVENTH PLAN REGARDING THE PROGRESS UNDER 20-POINT PROGRAMME

Year	Physical achievement (Area in million ha.)
1985-86	1.51
1986-87	1.76
1987-88	1.77
1988-89	1.90 upto 12/88.

Source : Report of the Working Group for Wastelands Development Sector in the Eighth Five Year Plan.

From the above table it seems that there is a substantial improvement in the rate of afforestation, though there is definitely scope for further intensification of work.

3.5.14 Schemes Proposed by NWDB

Various schemes and projects were initiated by NWDB in addition to those already ongoing in the name of social forestry.

Afforestation and Wasteland Development Schemes started in the Seventh Plan.

3.5.15 Rural Fuelwood Plantations

To meet the demand for fuelwood this scheme was introduced in the Sixth Plan. 157 districts were selected for fuelwood plantations. It involves raising of fast growing species, fodder and small timber both in block plantation and farm forestry. During the first three years of the seventh Plan an amount of Rs. 32.04 crores was spent and an area of 0.27 million ha. has been covered under the scheme. The farm forestry scheme has been implemented by the Forest Department in the States. It is centrally sponsered scheme, with 50% of the cost of plantation given as grant assistance to the States and 100% to the Union Territories.

3.5.16 Soil Watch

This scheme was introduced in 1977-78 for treatment of identified catchments in Micro-watersheds in the Himalayan States. This scheme involves mainly soil and moisture conservation measures, afforestation and pasture development and also treated selected micro-watershed in an integrated manner. During the first three years of the seventh Plan, an area of 0.10 millio hectares has been covered with an expenditure of Rs. 36.56 crores which was 26.42 crores in the Sixth Plan and 0.11 million hectare area was covered.

3.5.17 Grant-in-aid to Voluntary Agencies

This scheme was initiated in 1985-86 with a view to involve non-Government organisations in the process of development of wastelands with public participation. The main priorities of this scheme are—

(1) Establishmennt of decentralised nurseries and school nurseries; (2) Black plantations specially on community lands of SC/ST and people living below the poverty line; (3) Pasture Development through peoples institutions and peoples involvement; and (4) Assistance in implementation of the tree patta schemes.

The scheme is being implemented for the last three years. 248 projects involving 195 agencies were finalised upto 31st November, 1988 all over the country involving a financial commitmennt of Rs. 20.15 crores.

Seedligs raised in three years as follows.

Year	Seedlings.
1985-86	82 lakhs
1986-87	250 lakhs
1987-88	550 lakhs

3.5.18 Decentralised Peoples Nurseries

This scheme was initiated in the year 1986-87. To provide the planting material to local people, agencies, farmers etc., for income generation of local people by seedling production through farmers especially small and marginal and to establish nurseries.

Forest departmets, schools, NWDB have implemented this scheme as follows.

The Nurserymen were provided plants at the rate of 45 paise per plant.

(*a*) Kisan Nurseries have been set up in rural areas and 34.26 crore seedlings have been raised.

(*b*) School nurseries are being promoted on land made available by schools.

(*c*) Voluntary agencies which include Non-Government agencies are also playing important role in promoting these nurseries by involving landless, small and marginal farmers.

(*d*) NWDB with the help of Milk Union Federations have set up decentralised nurseries. The expenditure of Rs. 20.41 crores have been incurred on this programme. This proved that there is a large scale participation of the people and this scheme is useful in making people aware of the benefitis of afforestation. Certainly, this scheme helps in raising income of the poor people who get seedlngs at their door without any transport cost. With school nursery scheme, the new generation of school children become aware of the importance of trees.

3.5.19 Silvipasture Farms

To meet the severe problem of grazing and fodder this scheme was introduced in the year 1987-88 to augment the induction of good quality forests on Government community and marginal private lands. This will also help for encouraging stall feedinng. This is a simple scheme of regeneration of grasses and planting fodder trees. Being a centrally sponsored scheme 50% assistance is provided subject to a maximum of Rs. 1250 per hectare.

3.5.20 Externally Aided Social Forestry Projects

World Bank, USAID, SIDA, CIDR, ODA (UK) & DANIDA are giving financial assistance to Social Forestry Projects, which have been started in 14 States since 1981-82 for a period ranging from five to eight years and which envisaged tree planting and afforestation of 19,84,600 hectare of wasteland with total investment of Rs. 911.73 crores. These schemes are sponsored to fill the gap between demand and supply of fuelwood, fodder and small timber for rural population by tree planting and afforestation on wastelands, marginal farm lands and other vacant lands. The important activities like nursery raising, planting and protection of trees with people's participation through awareness raising about environmental and economic benefits of tree planting programme, are being implemented through Social Forestry Projects. This scheme also involves education, training, research and extension activities to develop technical and managerial skill of the State Government departments and individuals tree growing.

The success achieved through this scheme is noticeable but not satisfactory. The major fault is on the part of implementing agencies. This is the general experience in India that any scheme which is good on paper is either impracticable or its implementation is faulty.

3.5.21 Seed Development

In the ongoing programme of afforestation seeds being used are collected from all and areas which adversely affect the growth and vigour of the planted trees/shrubs. NWDB introduced a new central sector scheme in the year 1988-89 for seed development which include seed grading and distribution of standard seeds of required tree species, grasses and legumes.

3.5.22 Other Schemes of NWDB

Area oriented fuelwood and fodder projects (AOFFP), Aerial seedling programme, plantation of minor forest produce and margin money scheme are the new schemes which are initiated by NWDB from 1988-89.

AOFFP programme is proposed to be implemented in 52 districts of 11 States, with the help of Department of Agriculture and Co-operation to ensure integrated development of wastelands in the identified watersheds.

Aerial seeding programme is very effective. In remote, unaccessible areas like ravines, hills, etc., vast tracts are quickly covered under this scheme. Therefore, it is a cost saving scheme. Few States are selected for this scheme.

Plantation of Minor Forest Produce includes Bamboo, Fruit bearing species like Tamarund, Mahuwa, Mango and medicinal plants. 100% assistance is provided by the Central Government for implementation by State Governments.

'Margin money scheme' is a Central sector scheme. In this scheme 25% of the project cost is given as grant (margin) by the NWDB, provided an equal matching contribution is given by the eligible institution and State and at least 50% of the total project cost is financed by a financial institution. This schme is initiated to encourage the flow of institutional funds by which socially beneficial afforestation and wasteland development project can be carried out through active public involvement.

3.5.23 Rural Employment Scheme

To generate the rural employment State Forest Department have implemented afforestation component under NREP (National Rural Employment Programme) and RLEGP (Rural Landless Employment Guarantee Scheme) programme earmarked for Social Forestry component on Government community lands or road sides and canal embankments. Afforestation is also carried out on degraded forest lands in the absence of community land. The fund for this programme is made available by the Rural Employment programme. Evaluation carried out by the Programme Evaluation Organisation reveals the fact that people's participation is very minimal and it is mainly carried out by a Forest Department.

Farm forestry component which was mainly started for providing fuel and fodder to the rural people is now being executed mostly to meet the requirement of poles for urban markets.

Under soil conservation schemes "Integrated watershed management in the catchment of flood prone rivers" and soil conservation in the catchments of River Valley Projects programme were sponsored. During the period of Third Five Year Plan and Sixth Plan and now these schemes are being implemented in continuation.

3.6 Strategy for the Eighth Plan

Land degradation led to a serious Ecological and Socio-Economic crises. No doubt efforts have been made in the forestry sector though the efforts being made in the last forty years have not resulted in achieving the object of needed forest cover in the country. Inter-sectoral linkage among all development departments would also be necessary to make eco-development and conservation efforts of the Government, a success. Considering this a strategy the

Eighth Plan was formulated to focus on (*a*) halting degradation and (*b*) improving productivity on degraded lands. Environment and Ecology has to play an important role with co-operation and co-ordination of all development sectors and departments like agricultural research, rural development, water resources, non-conventional energy sources and sciences and technology. All these development departments would have to be integrated and involved directly in planning as well as implementation.

3.7 Approach to the Eighth Plan

Considering the value of restoring the Ecological balance as well as to meet the growing demand for fuelwood, fodder and timber in the country, the overall approach has to be multi-disciplinary and multi-pronged, based on integrated landuse planning on a watershed basis.

3.8 Objectives of the Eighth Plan

Eighth five year plan would have to incorporate the following main objectives.

(*a*) Preservation of biological and genetic diversity in terms of flora and fauna.

(*b*) Conservation and protection of the ecology, environmental and forest cover from further degradation due to encroachments, illicit fellings, fires, shifting cultivation, excessive grazing etc.

(*c*) Increasing the productivity of forests.

(*d*) Survey and demarcation of undemarcated forests.

(*e*) Practising scientific forestry on the principles of well prepared management plans.

(*f*) Rehabilitation of the degraded forests by afforestation.

(*g*) Meeting the requirement of local people for firewood, fodder and small timber, minor forest products etc.

(*h*) Surveying of forest resources by using latest techniques.

(*i*) Meeting the requirement of wood for industries and Government organisations consistent with the principles of conservation and forest management.

(*j*) Promoting forestry research on modern lines and developing forestry extension for transfer of technology in the field.

(*k*) Developing professional competence by undertaking schemes of forest education and training at all levels.

(*l*) Encouraging proper and efficient utilization of forest produce.

(*m*) Establishing the symbiotic relationship between forests and the local people particularly tribals and promote programmes for their welfare.

(*n*) Promoting plantation of species yielding minor forest produce on which the local population is dependent for their livelihood.

(*o*) Management of wildlife to maintain genetic continuity between artificially separated sub-sections of migrant wildlife.

(*p*) Developing linkages with other departments like agriculture, rural, welfare, tourism, industries etc. as well as with non-governmental organisations.

(*q*) Building landuse planning capability, specially at district and village levels, aimed at integrated use/management of wastelands.

(*r*) Regeneration of degraded forest areas.

(*s*) Grassland and fodder development.

(*t*) Production and conservation of fuelwood.

(*u*) Reclamation of special problem lands.

(*v*) Promotion of agro-forestry/Farm forestry.

(*w*) People's participation at all stages and a conscious effort to involve small and marginal farmers and agricultural labourers.

3.9 Programmes Proposed to be Included Under Eighth Five Year Plan

3.9.1 Production Forestry

According to the draft proposals for the Eighth Five Year Plan the main thrust of the future programme during the plan period would be to revitalise and strengthen the ongoing programmes and take up new programmes in regard to forests.

Programme of natural forest improvement and increasing production of fuelwood and commercial wood of general utility etc. would be initiated to ensure sustained supply of timber for national needs (defence, railways, industries etc.) and the essential requirement of firewood, fodder. Schemes proposed under this programme are as under :

3.9.1.1 *Schemes of Rehabilitating the degraded forests*

Ten million ha. area would be selected for augmenting plantation to rehabilitate degraded forests for restoration of ecology and environment and undertake fuelwood and commercial plantation in forest areas and improve natural forests along with protective measures. Outlay proposed is Rs. 10,000 crores, the Central assistance would be 100% to the State Government.

3.9.1.2 *Scheme of Intensive Management of Potentially Productive Forest Areas*

Programmes of intensive management of potentially productive forest areas in 50 lakh ha. to raise productivity of the forests through the introduction of fast growing species. Operations include complete fencing, soil working, soil and moisture conservation, improvement of existing staff and maintaining genetically superior species. The scheme also envisages protection of areas from grazing, fires and unauthorised fellings. This is a centrally sponsored scheme with 50% Central assistance from the Central Government and the balance can be arranged from the banks or from the States outlay.

3.9.1.3 *Scheme of Logging of Matured Plantation and Natural Forests*

Efficient application of modern logging techniques are needed in the harvesting and transportation of old and matured staff from plantation in forest areas. Efficient utilisation of wood by making use of modern techniques and minimising the wastage and improved seasoning and wood treatment methods to prolong life of the timber is also needed. Therefore, the scheme of the application of modern technique would be introduced in the Eighth Plan. In this scheme Central assistance would be 50% grant to the State Government and the balance can be arranged by the State from its own outlay or from banks.

3.9.2 Minor Forest Products

A major portion of Minor Forest Products *e.g.* plants yielding essential oils, medicinal plants, fatty oil seeds, gums, resins, dyes, fibres, bamboos, canes, tendu leaves, tusser silk, lac shellac etc. have been traditionally collected by the tribal people with or without processing and used either as food item, direct sustenance or traded in local market.

If the MFP resources are properly and systematically managed and their full potential tapped they would serve as an effective channel for socio-economic advancement of tribals and other rural people. Therefore, the scheme of MFP which is being implemented in the Eighth Plan needs to be continued with following important schemes.

3.9.2.1 *Important Schemes*

Plantation improvement of Bamboo forests. Bamboo is the important item of MFP for commercial promotion. Therefore, Bamboo plantation programme would be initiated in 2.5 lakh ha. area. This scheme covers improvement of bamboo forests through removal of dead/decayed clumps, working of soil, plantation of bamboo and use of advanced plantation methods. This is a centrally sponsored scheme with 50% assistance from Central Government.

3.9.2.2 *Plantation of Minor Forest Products Except Bamboo*

The main objective of this scheme is to increase the production of MFP in 50,000 ha. It is being implemented by the State Government and assisted by the Government of India. It envisages plantation of fruit trees, Tendu leaves, Harra, Behera and medicinal plants. During the Eighth Plan period the outlay proposed is Rs. 50 crores.

3.9.2.3 *Marketing of Minor Forest Products through Co-operatives*

The main thrust of this scheme is to introduce systematic and scientific methods of harvesting, collection, processing, grading and marketing of MFP. This will help the villagers and tribals to improve their efficiency and increase their income. It is a Centrally sponsored scheme with an outlay of Rs. 1000 crores, with 20% assistance to State Government The balance will be managed

Table 3.4
SCHEMES PROPOSED FOR THE EIGHTH FIVE YEAR PLAN – WILDLIFE

(Rs. in crores)

Schemes	Objectives	Activities	Outlay proposed
Ongoing schemes			
(*a*) *Centrally sponsored*			
(1) Development/Conservations of Natural Parks.	Encourage the States to set up more natural parks, and bring out overall improvement with infrastructure and financial assistance.	100% Central assistance for translocations of villages and new posts, 50% assistance for compensation of damages to crops/properties/human lives, and 100% assistance for works of habitat restoration.	100.00
(2) Development/Conservation of Sanctuaries	To afford protection to different wild biological communities and improve habitat management.	-do-	100.00

(3)	Project Tiger	To ensure maintenance of viable population of tiger and to preserve for all time areas of biological importance as a national heritage for the benefit, education and enjoyment of the people.	To increase the Tiger Reserve from 17 to 20, 100% central assistance for non-recurring new staff and 50% compensation for damage of crops and 100% assistance for ex-gratia grant of Rs. 20,000 per death and 10,000 for permanent disability.	40.00
(4)	Rehabilitation of Endangered species.	Support the programmes of captive breeding and preservation, where appropriate utilised species which are plentiful in one area for rehabilitation in another areas.	Construction of captive breeding enclosures and provision of equipment and research facilities and training.	5.00

(Contd.)

Schemes	Objectives	Activities	Outlay proposed
(5) Conservation Education and interpretation, programmes.	To promote conservation awareness among rural people living near the protected areas.	Set up forest and wildlife youth clubs and nature clubs and supporting their activities.	5.00
(6) Control over poaching and illegal trade in wildlife.	To afford protection to wildlife outside national parks and sanctuaries.	Setting up flying squads and anti-poaching squads to control poaching and illegal trace to protect species like Must Deer, etc. 50% assistance for staff and equipment and 100% for species pay.	10.00
(7) Development of Zoos on modern lines.	To assist and develop state owned zoos and set up a National authority to monitor and control the management of the zoos.	Development of infra-structure and provision of veterinary care, training of staff and captive breeding of endangered species.	24.50

(8)	Conservation of Rhinos.	Develop Rhino habitats in Assam, improve administrative structure, and strength of staff for affording protection to Rhinos.	100% Central assistance for non-recurring and recurring expenditure and provision of infrastructure for protection against floods.	5.00
	(b) Central Sector			
1.	National Zoological Park, New Delhi.	Development of the Park as modern zoo of National status.	Develop zoo as per the master plan.	10.00
2.	Directorate of Wildlife Preservation. (Strengthening of Wildlife Division)	The scheme will replace the existing scheme of Strengthening of Wildlife Division.	To administer all Centrally sponsored and Central Sector schemes, increase the number of Regional Offices from 4 to 8.	10.00

(Contd.)

Schemes	Objectives	Activities	Outlay proposed
3. Wildlife Institute of India.	It is autonomous National Centre for Training and Research in Wildlife. It will act as a nodal national agency.	Act as a Centre for Regional Training for countries in SAARC and South-East Asia, strengthening of training programmes for in service forest officers/ M.Sc. course and courses in wildlife extension.	10.00
New Schemes			
(*a*) *Centrally sponsored*			
(1) Eco-development around important protected areas.	To set up model areas of eco-development around selected national parks, tiger reserves and sanctuaries along the suggestions contained in the IBWL Task Force Report.	Selection of about 40 National Parks, Sanctuaries, Tiger Reserves, preparation of plans for eco-development *i.e.* pasture development cattle population reduction, breed improvement, alternate energy sources, agriculture, sericulture, horticulture development etc. (100% Central assistance)	60.00

(2)	Management Planning for protected areas	To implement systematic and scientific conservation programme after drawing of the appropriate management plans for each protected area depending upon its significance.	100% Central assistance in the preparation of management plan under a nodal officer under the department of wildlife preservation.	10.00
(3)	Project Snow Leopard	To strengthen the wilderness eco-system in high altitude Himalayas and Trans Himalayas.	100% Central assistance on recurring and non-recurring for comprehensive eco-system restoration and conservation in 10 to 15 selected parks/sanctuaries.	15.00
(4)	Project Elephant.	Conservation of Elephants with habitat fragmentation and ivory poaching, in protected areas not covered under project tiger.	100% Central assistance for recurring and non-recurring for the protected areas and corridor habitats between protected areas and major forest blocks.	20.00

(Contd.)

Schemes	Objectives	Activities	Outlay proposed
(5) Conservation of Marine Parks.	To prepare management plan for setting up Marine Parks because of fishing rights of local communities and illegal fishing by poachers.	100% Central assistance for sea-worthy petrol boats, other equipment and staff.	10.00
(6) Support for Wildlife Research and Training to State Wildlife Wings.	To promote wildlife research in protected area, monitoring of habitats and populations of crucial management.	100% Central assistance for staff, equipment, field work. One research centre for each Unit containing 50 parks/sanctuaries.	10.00
(7) Wildlife Staff and Welfare Centre.	To provide wildlife staff living in interior areas with welfare measures including free hostel and boarding facilities or allowance to mobile dispensaries, etc.	100% expenditure for facilities and special cash rewards for outstanding performance in wildlife protection, improvement and research.	5.00

(8) Establishment of forest parks and gardens in small towns/districts.	To provide assistance to State Governments in establishing forest parks and gardens in small towns near the forest areas.	50% Central assistance for materials including infrastructure/non-recurring.	10.00
(9) Development of infrastructure for the development of wildlife conservation.	To provide quicker information slow, a number of networks manned by dedicated personnels in all the sensitive areas in national parks/sanctuaries/Project Tiger reserves will be established.	100% Central assistance on non-recurring expenditure (wireless networks etc.) and 50% on recurring.	50.00

through banks and State Government's outlay by the State Government. This scheme will be implemented through the co-operatives.

3.9.3 Nature Conservation and Wildlife Management

Environmental security is maintained by two main aspects *i.e.* wildlife conservation and forest conservation. These two aspects together are essential for environmental security. Wildlife conservation is an integral aspect of forest conservation. Therefore integrated approach is needed to forest and wildlife management.

Local people depondancy on forest is the main problem common to forest and wildlife conservation. In November, 1983 the Government of India adopted a 10-point National Wildlife Action Plan (NWAP). It includes (*i*) Setting up of a biogeographically representative network of protected areas, (*ii*) Improving their management, (*iii*) Protection of Wildlife in multiple use areas, (*iv*) Training & research, (*v*) Captive breeding of endangered species and (*vi*) Education and interpretation. To facilitate the research and training for officers and workers, the Wildlife Institute of India was established in 1983. This boosted the training and research effort in wildlife conservation and management.

During the Seventh Plan total allocation (Central and State) under the wildlife sector was a meagre 7% within the forestry sector. Unless the Govt. strengthen imputs into the existing schemes and launches new schemes, desired objectives of wildlife conservation cannot be achieved. Schemewise details of Centrally sponsored and Central sector schemes are given in Table 3.4.

3.9.4 Forest Protection

The basic important issue is to protect the forest whatever forest and forest land is there. Forest protection involves two main objectives *i.e.*

(1) Protecting the forest from illicit felling, encroachment, fire damage, grazing etc.

(2) Reducing the damage the forests from insects/fungus/diseases.

Tabie 3.5

SCHEMES PROPOSED FOR THE EIGHT FIVE YEAR PLAN UNDER FIRST PROTECTIONS

Sr. No.	Scheme	Objectives	Activities	Outlay (Rs. in crores)
New Schemes				
1. *Centrally sponsored*				
(a)	Creation of protection Force and re-organisation of State Forest Deptt. for protection.	To ensure effective protection of forests from organised illicity felling and encroachments and unauthorised grazing and to strengthen State Forest Deptt. for the protection of forests.	For every 300 Sq.km. forest area one mobile squad consisting of one head guard, 3 Guards and 1 Range Officer alongwith Jeep and arms will be provided. Pattern of Central assistance 100% Non recurring and 50% recurring.	100.00

(Contd.)

Sr. No.	Scheme	Objectives	Activities	Outlay (Rs. in crores)
(b)	Development of Infra-structure for protection measures in forest areas.	To provide quicker information flow to the preventive force for the protection of forests.	Establish Wireless & Protection networks in all forest sensitive areas with infrastructure. Pattern. of assistance is 100% non-recurring.	40.00
(c)	Introduction of modern Forest Fire Control Methods.	To prevent, detect and suppress forest fires.	Construction and maintenance of fire lines, watch towards etc., and provision of 'Fare Finders' & Wireless equipment and public education covering 75,000 Ha. Pattern of assistance is 50 : 50 basis.	18.00
(d)	Provision of protection services against fungus/ insects/diseases.	To protect the forests from diseases.	Provide modern plant protection services.	2.00
			Total	160.00

To achieve these objectives, 10% forests area should be covered which is most sensitive to organised illicit felling and encroachment. Dedicated personnel at forest division level in sensitive forest areas are needed for protection.

Therefore, State Forest departments should be strengthened to reorganise the protection and administrative units, so that each unit is effectively protected by the staff against illicit felling, encroachment and fire. Diseases are always harmful for the growth of the forests. There should be a provision of protection services against diseases, fungal and insect damage.

The schemes proposed for Eighth Five year Plan under forest protection are given in Table 3.5.

3.9.5 Forestry Research, Education & Training

New and dynamic approach is essential to achieve the new goals in forestry research, education and training. Therefore, forestry research has been reorganised and the Indian Council of Forestry Research & Education (IGFRE) is established to develop an area of specialisation in research institutes besides working in common areas of forestry research of the region. To develop the feeling of love towards forestry sector, the universities should introduce forestry subject to their syllabus. Through National Demonstration Project, the "Lab to the Land" programme should be started. Forestry sector is considered the main sector which provides employment on large scale to the rural people. In spite of this, the sector can provide a generate employment to qualified and trained students which are created through forestry research, education and training.

The schemes which would be sponsored in the Seventh Plan are given in Table 3.6.

3.9.6 Resource Assessment and Planning

As stated in Chapter 2 the forest area is only 75.18 million ha. of the geographical area of the country. Out of this 66.65 ha. are under Government control and 8.53 million ha. privately owned. Of the Government controlled forests, reserve forests formed 86.79% in 1947-48, 69.92% in 1950-51 and 60.28% in 1986-87. The interpretation of the landsat imagery indicates that of 75.18 million ha. contained actual forest cover and of this actual cover, only 35.77

Table 3.6

SCHÈMES SPONSERED IN THE EIGHTH FIVE YEAR PLAN

Sr. No.	Scheme	Objectives	Activities	Outlay (Rs. in crores).
1. Ongoing Schemes (Central Sector)				
	(*a*) *Farestry Research.*	Six new institutes have been structured discipline-wise. These will be addressing themselves to the priority research areas relevant to each zone in which these are located. Not only larger funds should flow to research activities but also a quantitatively larger and qualitatively superior manpower will be available for pursuing various research projects in the right direction.	Except FRI Dehradun other institutes are presently going through the phase of establish-ment. Construction of infrastructure and recruit-ment of suitable personnel is to be expedited. Model established by green revolution in agriculture needs to be replicated with suitable adoptions due to the longer gestation period and large number of species involved in forestry research.	95.00

(*b*) *Forestry Education*.	To improve the overall standards of forestry science in the country, forestry education needs to be imparted in the Universities without proliferation of substandard academic programmes.	To provide support to forestry education in the State Universities. Fourteen Universities have already set up separate departments/faculty of forestry.	10.00
(*c*) *Forestry Training*.	To impart training at five levels *viz*. Indian Forest Service, State Forest Service, Forest Rangers, Foresters and Forest Guards. Training of IFS Officers, and SFS Officers is looked after by Central Govermen. and others. To regroup States/ establishment of New SFS colleges	Annual training, capacity of the IGNFA is about 150 officers, 3 State Forest Service officers, Colleges 200 and Rangers colleges 700. It is proposed that Rangers colleges at Coimbatore, Balaghat, Chandrapur and Kurseon should continue to be run under the Government of India.	

(*Contd.*)

Sr. No.	Scheme	Objectives	Activities	Outlay (Rs. in crores)
		in view of difficulties in imparting training through a common language, to redraft the syllabii and to continue the Ranger colleges instead of closing down or transferring to State Governments and open new Rangers colleges.	Six Regional Forest Rangers Colleges at Haldwani, Rajpipla, Angul, Deomali, Chandrapur and Coimbatore be established/ continued. The outlays are:	
			(i) Indira Gandhi National Forest Academy.	3.00
			(ii) SFS Colleges (3 Nos.)	6.00
			(iii) GOI Rangers Colleges (4 Nos.)	5.00
			(iv) Rangers Colleges (State Governments) and Forests Guards Schools	6.00
				20.00

2.	**New Schemes—Forestry Research :**				
	(*a*)	Funding of Research through Universities.	To support forestry research particularly in agro-forestry systems in State Agri. Universities.	To fund worthwhile schemes in State Agriculture Universities.	15.00
	(*b*)	Funding of Research through Institutions/ NGGs./State Government	To support large number of autonomous institutions/ NGOs in forestry programmes.	ICFRE may fund these institutions and also joint sector research institutes like Indian Plywood Research Institute (IPIRI), Bangalore for research of forest.	30.00
	(*c*)	National Demonstration Project.	To set up National Demonstration Project under "Lab to the Land" programme.	States should be asked to set up research-cum-demonstration areas where forest species can be used for amelioration of sites and increasing productivity of wastelands.	10.00
	(*d*)	Centre for Manpower Development.	To train Scientists and Forest Officers selected for forestry research in research methodology.	To set up a centre at the ICFRE headquarters, Dehra Dun.	5.00

(Contd.)

Sr. No.	Scheme	Objectives	Activities	Outlay (Rs. in Crores).
(*e*)	Strengthening of Extension Wing in the Research Institutes.	To create a full fledged extension wing in each institute to transfer the research findings to end users.	Strengthening of extension set-up in forest research institutes.	10.00
(*f*)	Forestry Research Information Service	To create an independent Directorate of Forest Research Information Service (FRIS) at ICFRE, Dehra Dun and collect information from State agencies/universities etc.	Organising computer based information service at ICFRE with sub-centres in six research stations and compiling data for the use of end-users.	10.00
(*g*)	All India Co-ordinated Research Projects.	To carry out coordinated research on damage to forests due to heavy biotic pressures including fires, grazing, fuelwood an fodder (tree species).	Provide special grant for procurement of equipment and recurring expenditure through ICFRE to the scientist from the nodal institute to carry out the research in specified areas.	10.00

(*h*)	Training Programme (Mid career)	To start training programme for the middle and senior level officers in order to keep the field level officers with the latest trends in policy formulation as well as technological advances and creation of special cell for training.	To conduct junior level programmes with specialization (6-9 months), middle level programmes for Conservator of Forests (3-4 weeks), senior level programmes to Chief Conservator of Forests (one week). Top level programmes to Principal Chief Conservator of Forests, Orientation course for Range Forest Officers at the time of promotion of SFS, IFS and compulsory training of IFS Officers in Management Development programme.	5.00

million ha. could be classed as close *i.e.* with a crown density of more than 40%. To maintain the ecological balance and preservation of environment, the New forest policy document, 1988 aims at 2/3rd Land under forest cover.

Proper planning is needed for available resource assessment, to achieve the objectives of increasing the land under forest cover. The correct information about available forest resource can be available only through the preparation of comprehensive state of forest report which includes National vegetation map once in every two years. Under preparation of forest inventory work, survey to identify unsurveyed and degraded undemarcated forest lands *i.e.* 13.27 million ha. should be reported. Application of modern forest techniques strengthens general management education and training Schemes included under the Eighth Plan.

3.9.6.1 *Forest Survey of India*

1(A) Ongoing scheme of Forestry Survey of India was reorganised from 1.7.1986 which includes comprehensive state of forest report to collect, store and retrieve necessary forestry and forestry related data. This scheme also emphasises on forestry inventory in selected States to design methodologies relating to forestry survey, etc. The activities include the matic-mapping, vegetation mapping, electronic data processing, inventory of forestry, resources in north-eastern region, training of personnel in modern techniques of forest surveys, etc.

3.9.6.2 *Application of Remote Sensing Technique*

(B) Ongoing scheme of the application of remote sensing techniques include the assessment of the extent of the forest cover, classification of the forest cover by density classes, training staff in the application of remote sensing techniques, computerised data base management and forest inventory etc.

3.9.6.3 *Indian Forest Resources Information System*

To achieve this objective National Data Management Centre was settled which will be provided a VZX-11/780 computer system. Installation of this computer will be completed in 1989-90.

3.9.6.4 *Other Schemes*

New schemes include survey demarcation of forest land which require an outlay of 350 crores at an average cost of Rs. 250 per ha.

including erection of boundary pillars for surveying and demarcating the 13.27 million ha. of degraded and undemarcated forest land. This land will also be taken up for surveying the capability of land. Gradually the entire 75.18 million ha. forest land will be covered under such a survey. In Eighth Plan under new schemes the emphasis will be on forest inventory management which include strengthening the Indian Institute of Forest Management, Forest Resources Survey of India, Forest Resources Information System which will be able to provide smooth flow of information from one level to another.

3.9.7 Tribal Welfare

As per 1981 census there are about 538 million Scheduled Tribes in the country. It is estimated that most of the nearly 48 million forest dwellers living within or in the vicinity of forests are tribals. The tribals are well versed with the knowledge of forests, its treasures and bounties. They depend upon forests for sustenance.

The national forest policy 1988 lays greater stress on involvement of tribals in the protection and regeneration and development of forests and they should also be provided employment. For this forest management should associate the tribal people closely. While proposing and implementing the ongoing schemes and new shemes, it is essential to concentrate on issues which are practical and realistic in handling of tribal development through forestry and weightage should also be given to the objectives laid in forests. The objectives are—

(1) To provide minimum needs to people living within the forest villages.

(2) To provide minimum amenities and facilities to people living in villages in and around the forests and not termed forest villages. To establish training centres for augumenting the skills in people to enable them to earn higher income and increase their efficiency.

3.9.8 Schemes Proposed for the Eighth Plan

New schemes

3.9.8.1 *Programme for development of forest villages*

Human settlements in forest villages to eliminate poor living condition of the people living in these forest villages is needed to provide basic needs and amenities. Outlay proposed for this scheme Rs. 130 crores.

3.9.8.2 *Development of Infrastructure for Education, Drinking Water, Health, Roads etc., in Villages near Forests*

The aim is to improve their living by making provision of education, drinking water, health, roads, etc. Outlay proposed is Rs. 120 crores. The Central assistance would be 100% grant to State Government.

3.9.8.3 *Establishment of Vocational Training Centres*

Through vocational training centres, extension services will be provided to tribals. The outlay proposed is Rs. 20 crores with 100% Central assistance to the States.

3.9.8.4 *Control of Shifting Cultivation and Proper Landuse in Tribal Areas*

Resettlement of people practising shifting cultivation outside the forest areas and to motivate them for permanent cultivation and proper land use should be done by covering 3000 families every year. The target is about to 15000 families. Total outlay proposed is Rs. 40 crores, with 100% assistance of Central Government.

3.9.9 Other Schemes

3.9.9.1 *Strengthening of Forestry Division/Regional Offices*

The Five regional offices are established at Bhopal, Bhubaneswar, Lucknow and Shillong as part of the Secretariat of the department to deal with forest conservation matters. The outlay for the Eight Plan period is Rs. 6 crores. One more office will be opened at Chandigarh in view of the increasing work.

3.9.9.2 *Creation of Central Minor Forest Produce Division in the Ministry*

Minor Forest Produce division scheme will be sponsored by Central sector. This division will take care of all aspects relating to minor forest produce in the country. The outlay proposed for the Eighth Five Year Plan is Rs. 20 lakhs.

Table 3.7
PHYSICAL TARGETS AND PROPOSED OUTLAY FOR THE EIGHTH FIVE YEAR PLAN

Sr. No.	Programme	Unit	Physical	Proposed outlay (Rs. in crores)
1.	Production Forestry Plantation.	Million Ha.	10.50	11000.00
2.	Minor Forest Products Plantation.	Million Ha.	0.30	1175.00
3.	Nature Conservation and Wildlife.	-	-	500.00
4.	Forest Protection			
	Forest Fire Control sq.km.	-	75000.00	160.00
5.	Forest Research, Education and Training.	-	-	220.00
6.	Resource Assessment & Planning			
	Survey and Demarcation.	Million Ha.	13.27	
	Assessment of land capacity.	-do-	13.27	581.00
7.	Tribal Welfare.	-	-	310.00
8.	Others.			
	(*i*) Strengthening of Forestry Div./Regional Offices.	-	-	6.00
	(*ii*) Creation of Central Minor Forest Produce Div. in the Ministry	-	-	0.20
	(*iii*) Creation of Tribal Cell in the Ministry.	-	-	0.30
			Total :	13952.50

3.9.9.3 *Creation of Tribal Cell in the Ministry*

Tribal cell division will take care of all aspects relating to tribals, scheduled castes and scheduled tribes, tribal sub-plan, plan for SC/STs in the country. The outlay for the Eighth Plan period is Rs. 20 lakhs.

3.9.10 Physical Targets & proposed outlay for the Eighth Five Year Plan.

The proposals for the Eighth Plan have not yet been finalised. However, the tentative allocation of outlays for different programmes under forestry and wildlife are given in Table 3.7.

CHAPTER IV

ECOSYSTEM : ENVIRONMENT, FORESTRY & SOCIAL FORESTRY

4.1. The Ecosystem

The term ecology is derived from the Greek word "Oikos" meaning "house". The ecology, therefore, means the study of the house, the house of man *i.e.* earth with the plants, annimals and microorganism which live together on it with man. Ecosystem, Environment and human being are considered as the parts of same system. To know close relation beteween Ecosystem, Environment and human being, the Ecosystem model is given below.

4.2 Ecosystem model

In an ecosystem every organism has a specific role to play Eco-system has five components :

(1) A source of energy (E)

(2) A source of Chemicals (C)

(3) Mechanism for using the energy and materials, may be living or non-living

(4) Flow pathways indicating energy flows

(5) Interactions where, energy, material and mechanism interact to modify, or control flows.

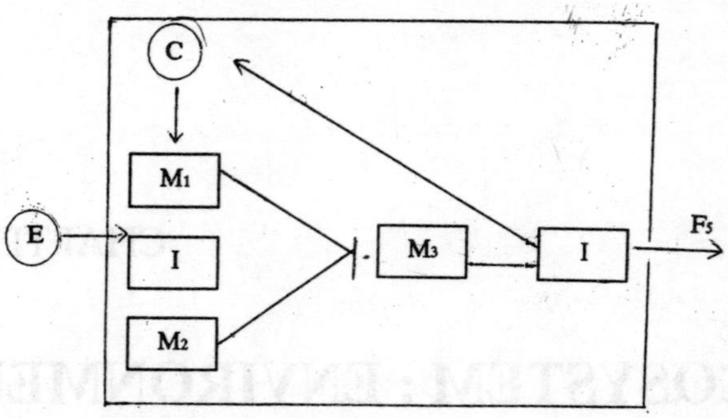

E—Source of Energy—Sun
C—Source of Chemicals—Soil & Air
M1—Plants
M2—Animals
M3—Microorganisms
I—Interactions
F—Energy flow or material transfer.

Source : Environmental conservation, J.B.Lal.

Sun is the chief source of energy by which the Eco-system exists and the life processes of the organsums are assisted. Plants directly absorb energy to make food with the help of soil and water by transferring this absorbed energy —Animals to obtain food from plants and Microorganisms from the dead plants and animals. This constitutes a food chain. Food chain is nothing but a sequence through which energy is transferred from one organnism to other. An Ecosystem comprises a series of inter woven food chains, which is known as food web .

Food Chain

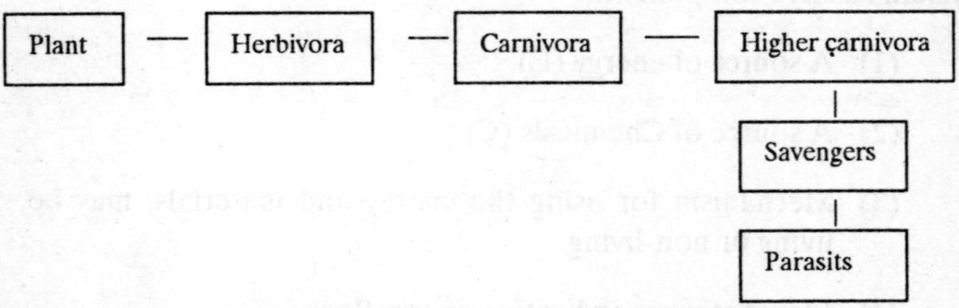

Law of Thermodynamics is also applicable to plants. Accordingly when any transfer of energy takes place some energy always escapes from the system usually as heat. Out of the energy received by plants less that 1% is transferred to men and animals. Thus it seems that the major role is played by plants *i.e.* receiving sunlight energy and transferring it to animal and man. The life of man and animals is solely dependent on number of green plants. These green plants are limited by the amount of sun-light energy and the efficiency of plants in converting it into chemical energy. It follows that in an ecosystem, the number of plants will be more than animals.

4.3 Inter-relations of Natural Resources

Eco-system, requires supply of chemicals which are obtained from soil atmosphere by plants which is an integral part of eco-system also including the inter-relations of natural resources in which plants are prime factor. This inter-relation is explained in the figure below :

ECOSYSTEM CONSERVATION

The interrelations of Natural resources

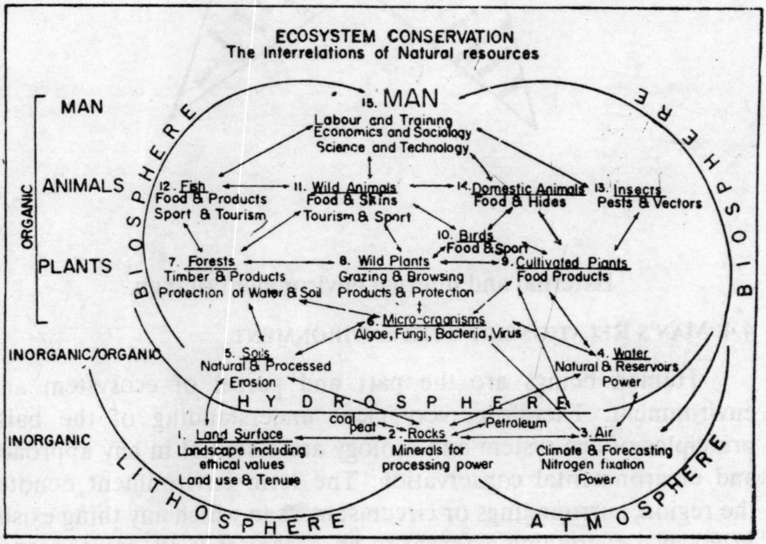

Source : Environment Concerns and Strategies.

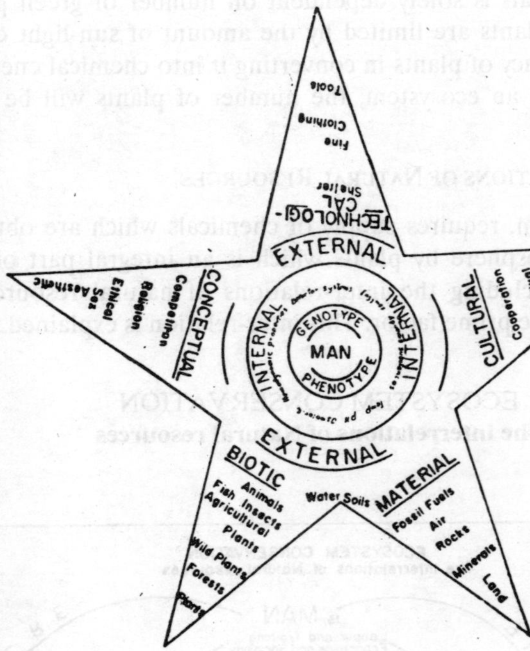

External and internal environment of man.

4.4. MAN'S RELATIONSHIP WITH ENVIRONMENT

Human beings are the part and parcel of ecosystem and environment. Therefore, very clear understanding of the baisc principles of eco-system and ecology are essential in any approach and environmental conservation. The term environment denotes the region, surroundings or circumstances in which any thing exists. In essence everything external to an organism is its environment. The organism in turn is the part of the environment of other organism.

The man as one of the several organims is reacting with the environment in his day-to-day functioning. He has to have close relationship with the society and natue by utilising various technologies. He is utilising the natural resources for his benefit and comforts by using the knowledge he has gained so far. There is a continuing quest for the balance and harmony between man and nature and the rational use of the planetary resources. Thus the man, with variety of functioning for comfortable living is affecting the environment not only of himself but also of other organisms prevailing in the environment and ecosystem. He has to be very careful about it and should attempt for the conservation of environment. He should also be aware of the expanding cosmos and its effect on natural resource development. He has to use his sense, creativity, different arts, culture for environmental understanding.

4.5 Environmental Conservation

The environment of man includes the non-living factors like land, water, atmosphere, climate and living factors like animals, plants, bacteria and viruses. There is always a co-existence and inter-dependence amongst living factors and non-living factors maintaining the balance or equilibrium in the whole system. Environmental conservation implies the implementation of such wise principles and measures whcih would lead to the maintenance of the ecological balance. The man always wants to live in the best possible conditions. This is possible only when his attitude towards his surroundings and nature is sympathetic. This conservation of environment does not mean only saving of forest flora and fauna, but is something more.

In the words of Ian Wilson, the former Australian minister for Home Affairs & Environment, "Conservation is not just about our forests, or native flora and fauna or our primary industries, important though these things obviously are. It is about the way we live and the way we behave whether it be as Governments, as local government authorities, as companies and industrial concerns or as individuals." Conservation is a positive and constructive way in which we can make sure not only that we make less of a mess but also that we leave the world not just as we found it, but in an even better condition for our children and generation to follow. Therefore, environmental conservation aims at preserving and developing the surroundings of man which includes living and non living factors by making the best use of nature for the welfare of the present as well as the future generation.

Patrick Jenkin, the former British Secretary of State for the Environment said, "Conservation is a paradox. People want to preserve the plants and animals but they also demand food, fuel and water. For food, fuel and water people need farms, power stations and reservoirs involves sacrifice of forests and plants and animals that live therein. Conservation is not an easy process. Conservation means striking a balance between the economic need for modifying the nature and the environmental necessity for keeping it as close to the original form as possible. The advancement of science and technology of the modern era has increased the level of man's economic needs to sky. The needs of the man are not limited to the basic necessities only. He is encroaching on environment to fulfil his desires. Therefore, to strike the balance between the satisfaction of man's needs and conservation of environment is the most difficult task like walking on a light rope. It is a false thinking that what is natural is "good" and what is not natural is "bad". Present era in which more stress is on well being of human beings, needs economic growth at higher rate and environmental stability also.

In the past, India, being underdeveloped country, more concentration was on economic development activities to achieve the present developing stage which led to the environmental damages and the misuse of nature. Man became blind in his efforts to maximise the material benefits. He has ignored the fact that material benefits cannot be achieved without environmental stability. No doubt man has to modify nature as per his requirements to increase the economic growth. But at the same time he should not destroy the natural system on which he depends. Environmental conservation requires a wise choice which will enable him to live in the best attainable world. Without disturbing the natural system, modification in nature will lead to the achievement of dual goals *i.e.* Economic growth and Environmental stabaility.

4.6 Concern in Developing Countries

The different approach towards country's development should be eco-system specific. Developmental works should not disturb the traditional but use system and must involve people. People's involvement in development could be achieved when the village is taken as the unit for development. Creation of powerful village institutions, funds, suitable changes in existing forest and revenue laws and a coherent village plan form the right development strategy. Government can set up village funds for diffferent development works instead of spending on various anti-poverty programmes.

Developing countries need fast growth to achieve their successive development stages in order to meet the basic essentials for life of its teening millions. All development requires a resouce based economy and it takes place in the background of a given physical and socio-economic environment. Resource based economy always exploits natural wealth and it alters environmental fabric. But present era calls for the new strategy which will be aimed at protecting the resource, or environment for sustaining development. Optimum use of natural resources is required for attaining the maximum growth. Optimum use always leads towards maximum benefits or profits. The past has shown the consequences of the mis management of the environmental resource, be it a forest, stream, land or coast. Therefore, new strategy is needed in Environmental management.

4.7 Environmental Management Strategy

In a planned and controlled economy, Government is a supreme authority in allocation of the resources for development activities. Therefore, it can be said that Government is thus the supreme manager of the environment. The main object of the Government in socialistic pattern is the "Maximum welfare of the maximum people". This is concerned with wise management and optimum use of resources. Government carries out the function of environmental management when it makes efforts for maximising the agriculture production, using the energy resources for energy production and distribution and while defending the nation against aggression. The best environmental management should aim : Conserve what needs to be conserved, protect what needs to be protected and regulate utilisation of natural resources to acceptable limits.

In a country like India, Management strategy of environment, calls for rational choice in which potential of environment should be assessed, because environment supports life system as well as it provides resources for development. The exploitation of resources for development process should not be at the cost of destruction of life and system. Therefore, management of the environment should think of the capacity of the environment.

4.8 Environmental Management Plans

Best Management Plan includes prevention, protection and correction. It is said that prevention is better than cure. In development process the problem of pollution becomes always severe. To avoid the pollution policy of decentralization of industries should be adopted. Choice of area to establish the various industries is an important factor in development process. Therefore, prevention requires the choice of proper area for the establishment of Industrial sector. Agriculture sector is not also free from creating pollution. Too much use of insecticides is harmful and is creating the disequilibrium in ecosystem. The expert technicians can implement the proper strategy to avoid the excess use of chemicals and fertilisers in agriculture which will prevent the further pollution and destruction of fertile surface of the agriculture soil. The electrification is always harmful to the forests. This can also be prevented by extending the electric line as far as possible out of the forest. Similarly irrigation projects involving forest land for submergence need to be weighed properly with its impact on environment.

The main and severe problem which is responsible for the ecological imbalances is the unlimited population growth. With the implementation of intensive family planning programme population gorwth can be checked. Thus, it is clear that there should be a limit to the population that can support the ecosystem and it would be unwise to grow beyond this limit. Population regulation is a concomitant necessity for environmental management of urban area. A very similar problem is that of management of urban area. In the present industrialisation era., the big cities are over-crowded and they are growing beyond optimum size. Supply of essential services also become impossible and consequently the quality of life is being deteriorated. To maintain the desirable standard of life of human, the growth of population needs to be checked. The Environmental elements are, a few kilometres of air without which nothing can survive. These elements should not be depleted and eroded and polluted. The present trends show that the life of human being can slowly become extinct because of high rate of pollution and erosion. To prevent such doom, environmental consideration in the management practices of life supporting systems must be adopted now.

Like prevention another important aspect of management practises is protection of environment which aims at conserving the available resources from degradation. Protection implies the wise use of resources in the environment for sustained development. In no way conservation and development are contrary. They are equally necessary for the continued survival of living systems. Social and economic gains should not be considered at the cost of renewable resources but these should be done in an ecologically sound manner. Protection needs maintenance of genetic diversity, combating soil degradation, controlling deforestation and overgrazing, arresting the spread of deserts, curbing exploitation of renewable and non- renewable resources and abating environmental pollution. These are all the prime aspects of sound environmental management. It can be described as, "Sound environmental management is the optimal allocation of limited resources between different possible uses. Environmental criteria and Economic consideration favour that such allocation should be efficient. Simultaneously, the available resources should be protected from degradation and scarce and diminishing resources should be conserved."

The third and important task is the implementation of corrective measures. At present whatever damage is being done to the ecosystem and over-exploitation of natural resources should be controlled and the resultant loss if any should be regained through the implementation of various schemes, such as National wasteland development programme, Afforestation scheme, social forestry, farm forestry etc. The forest land which is diverted to industries, can be coverd with trees through the implementation of green belt scheme, which is included in the guidelines of the Government in the environmental clearance.

The cornerstones of environmental management include, Environmental planning, Environmental status evaluation,

Environmental impact assessment and Environmental legislation and administration planning.

In a developing country, where the resources are lmited and objectives are too plenty, Environmental planning programme needs an integration approach of environmental considerations in the economic development planning process. This approach should be flexible as per requirements of the planned economy. This also aims at socially desirable programmes of development with least damage to ecosystem and needs also a concious view of long-term objects in preference to the more short-term economic goals. The modern era wants new social and economic objectives to be set because unstable economy and increasingly complex society link between population growth, resource use and development goals will lead towards achievements of the object of sustained development. More global planning porcess is also needed which is continuous, integrated and environmentally sound. Dynamic balance between environmental, social, political and economic concerns nessitates a conscious policy framework.

4.9 Environmental Impact Assessment

Ennvironmental impact assessmennt is a procedure useful for the environmental manager and decision maker to assess the potential efffects of the humann activities on the environment. The quantification of the effects of pollution is rather easier. However, it is very diffficult in respect of long-term ecological and social impacts. In such cases impact is assessed by comparison of the development options or the comparison of the same option at different sites with the sole object of conserving the environment.

The environmental impact assessment is a recent technology and the procedure that is now available is not yet capable for clear conclusions. The conclusions are based on the value judgement. This is mainly because the standards for the different environmental values have not been finalised.

A procedure of combining environmental and economic consideration on the basis of depletion of resources is now in vogue. This is nothing but an extension of cost benefit analysis ascribing monetary values of the environmental factors affected and monetary cost to restore them.

Because of the difficulty of assessment and subjective judgment environmental impact assessment is criticised. However, such studies are found useful and further research and studies would refine the procedure which will be of practical importance.

4.10 Control of Pollution

As discussed above the increasing pollution leads to the severe environment problem. Therefore, there is a need of preventive approach based on management practices and by creating awareness to minimise pollution. Pollution prevention includes the following measures.

(*a*) *Clean techologies*. Clean technologies aim at maximising the conversion of raw material into products and minimising the quantity of fuel and water used. This implies the recycling and reuse of water and utilisation of treated waste water from industrial units and municipal areas for irrigation. Adoption of such measures should be encouraged.

(*b*) *Establishment of new industries*. Micro level plane detailed maps will have to be prepared for demarcating the prohibited areas which are already covered with many industries. New industries should be set up only when they are environmentally sustainable. This is one of the most essential factors like transportation power. Raw materials should also be considered for locating the industry. The proposal of new industry should be accompanied with environmental impact assessment without which no permission for setting of industry should be given. This study includes information about hazards risk analysis of all units operations and processes, a disaster preparedness plan, appropriate considerations on concentration of other hazardous industries in the same locality and continuous monitoring programme in the overall planning of industries functions.

(*c*) Setting up environmentally balanced industrial complexes in which the waste from one industry can be utilised by other industry should be considered. Otherwise it will lead in creating pollution.

(*d*) Growth centres and cluster of small scale industries should imply an environmental impact study. Such studies be encouraged and designed by the authorities.

(*e*) State house containing hazarduous chemicals should not be located near rivers of canals so that in case of any mishap, the chemicals do not get washed into the water system.

(*f*) Appropirate legal framework should be created for all the factors mentioned above.

Only Laws and formulation of policies cannot minimise the pollution effect. Social awareness is the important factor. The intensity the improvement of activities of reducing pollution on creation of public awarness in agencies which cause pollution (industrial units), and in public who suffer from pollution is needed. To achieve the objective of creating public awareness the continuous programme of educating the people through school and college curricula and also through the education of factory workers is needed. This programme is being implemented in some States and to some extent, it is successful also. Rewards and incentives for people and industry should also be encouraged.

4.11 Publicity for Environmental Awareness

Mass media and Television are the best instruments to achieve public awareness for pollution. Numerous agencies of the government ranging from the departments of Environment, youths, women, educationists and personnel to the National Wastelands Development Board and the National Council for Educational Research and Training are also doing efforts of launnching programmes regarding the public awareness for environment conservation. Voluntary agencies are also involved in this work. Any difficult work can only be successful when co-operation from public on large scale is received by implementation agencies. The public awareness regarding environmental conservation is only limited to wildlife protection and protection of forests. But there should be comprehensive thinking which includes all components of Ecology.

4.12 Environmental Status Evaluation

On the basis of some parameters, it is possible to evaluate the environmental status. The success of implementing policies regarding environment conservation can be measured with the help of environmental status. Air, Wasteland, Flora and Fauna are the main components of Eco-system. For the evaluation of environmental status on microlevel we can observe the air quality, water quality, soil characteristics, productivity and population size, population dynamo, genetic diversity which are supposed the status index of air, water, land flora and fauna.

The another macro index of the environmental status can be indicated in terms of resource measurements. Estimation of available resources such as mineral reserves, fuel reserves, timber reserves in forests and other economically exploitatble natural resources is made for the purpose of assessing reliability of such figure, even approximately is useful to observe the environmental status. The reducing of resource potential through use and abuse can also be used to measure the present environmental positions. While doing observation of environmental status, it is essential to compare the status between two points in time or space.

In development process some visible material gains are acquired and quickly at the cost of invisible potential losses. Visible material gains can easily be measured in terms of many, which are in short term appealing to human beings but potential losses cannot be easily measured and are in terms of life and death of a species and for a system. These log terms losses also cannot be compensated by the probable gains. So environmental management aims at balancing among probable gains against potential lossess. The policy measures are always under heavy and multiple pressures because of the inter-linkages among ecological and socio-economic sectors of society. Therefore Administrators' inter-disciplinary skills to develop policies and to implement action programmes would ensure the potentially desirable quality of the environment.

4.13. Role of Forests in the Environment Management

The existence of the forests in desirable proportion helps maintain the environmental balance by checking soil erosion, improving the quality and quantity of water, ameliorating climate and checking pollution. However, the forest area recognised as reserved forests and protected forests is very meagre and is in charge of the Government. The percentage is already less than what has been prescribed by the National Forestry Policy. Whatever forests exist is under tremendous pressure and is therefore getting depleted and degraded day by day. Various causative factors that are responsible for the degradation and loss of forests cover are as under.

4.14. Causes of Depletion of Forests

The following factors are prominent.

4.14.1 Huge Population

The human population which was 35 crores at the time of independence, crossed more than 80 crores in the year 1990. Increasing population led to the shortage of food and unemployment also and exploited the natural forests at very higher rate. At the same time population failed to conserve and to increase the growth of forests. To solve the problems of food more and more forest lands have been cleared and utilised for agriculture crop. About a few years ago clearing the forests for agriculture was an established government policy of many developing countries. India is one of them. Today most of the developing countries are undergoing process of landuse change, under which forests are giving way to agriculture. Some forest soils may be most suited for agriculture but once the valuable trees are felled, their fertility is likely to fall down rapidly as nutrients tied up with the vegetative layers are leached away and severe soil erosion sets in. In some cases, highly productive forest land may be transformed into waste land within a short span of a few years. The confugeration of the land, economic condition and the soil's fertility are the main factor for deciding the utilisation pattern of a particular land. Unfortunately this important aspect has not been considered by Government. In some hilly area fertility is poor and land is shallow and it is very difficult task to convert such infertile and shallow soil into fertile land. Collosal efforts and huge expenditure are also in vain for soil improvement. It is very unsound economic proposition to take agriculture crop on such land which are best suited for regeneration of forest species.

4.14.2. Encroachment of Forests

This problem is probably worse than those caused by clearing of forests for agriculture on scientific lines. Moreover, because of various socio-political factors, this is one of the most difficult one to control. Most forest encroachments involve the clearing of trees. Construction of roads also have led to encroachments on forest roads, there is rush of settlers along these roads penetrating into the forests. Once this process starts, large number of families settle along the road side forest areas and they secure the largest possible forest area for themselves. These encroachers are from the lowest sections of society and have neither the skill nor the patience to manage the land on scientific lines.

4.14.3. Grazing

Grazing by limited animals does not cause much damage. But when a large number of animals enter the forest for grazing then there is an immense damage to trees through the excessive removal of foliage, consequently resulting in stunted growth. Animals hamper natural regeneration by eating or trampling young shoots and seed lines. Excessive grazing compacts the forest soil and affects the regeneration process.

4.14.4 Forest fires

In a short span of time forest wealth worth lakhs of rupees is lost. These forest fires damage the crops, regeneration, productivity of the forest, protective powers of forest, soil fertility and soil erosion, wildlife and aesthetic view.

4.14.5 Diversion of Forest land for Non-forest Use

Large proportion of the forest area has been lost as the forest land was diverted for the construction of dams, industries, construction of roads and for agriculture by regularising the encroachments. Of course this process has been considerably checked with the passing of Forest (Conservation) Act 1980 by the Government of India. No forest land can now be diverted without the prior permission of the Government of India.

4.14.6 Depletion of forests Vs Increase in Demand of Forest Produce

On one hand the existing forests are depleting causing immense environmental degradation, on the other hand the demand on the forest produce like timber, firewood and fodder is increasing day by day with the increase in the human as well as cattle population. Even if all the forest areas are made well stocked, it is not possible to meet the growing demand of the local rural population. Therefore, the remedy is not merely development of the existing forests for its productivity but to bring more and more area under forestry or tree cover which will not only take care of the environment but also meet the demand of the forest produce.

4.15 Energy and Forest Produce Needs of People

Discussing in detail about the wood energy famine in the country Plant (1984) states, "in India 95% of fuel needed for demostic consumption in rural areas, comprises firewood, cowdung, and agricultural wastes, The villager/farmer although fully aware of the importance of cowdung and agricultural residue as manure is compelled to use large quantities of dung as fuel due to shortage of fuelwood. Unless, therefore, wood resources are adequately created locally, it would not be possible to divert animal and vegetable wastes to the farms as manure. The author estimates that in India over 458 million tonnes of wet dung is diverted annually to the rural hearths. Assuming an additional yield of five quintals of foodgrains per hectare, the annual loss of food is thus estimated at 45 million tonnes valued at about Rs. 90,000 million." Like burning of dung for energy, removal of fodder from the forests has also worsened the problem of depletion of forests. It is a habit of the people residing the forest to release the cattles and sheep in the forests by which not only the forest produce is reduced but also damages the further growth and regeneration. A conservative estimate made by Acharya *et al.* in 1973 put the most of feed of a stall fed buffalo at Rs. 453 and they gave a figure of Rs. 37 crores as the total cost of fodder for feeding the entire country's domestic animals for a year. This estimate was made in 1973. To-day this figure must certainly have crossed the Rs. 100 crore mark. This clears the fact that there is a need to raise fodder plantations in non-traditional forest areas.

4.16 Social Forestry Concept

Large scale deplection of forests, increase in demand of the forest produce, and the consequent reduction in the supply of rural energy needs, have made the planners and administrators to think of a programme called social forestry. In fact, the concept was first discussed in the silvicultural conference held in 1946. Need for augmenting the forest or tree cover not only on the traditional forest areas but also on the Government wastelands, private lands not fit for renumerative agriculture, planting of trees on bunds, road side canal banks, railway line and the lands in charge of other Government department was stressed. This would not only take care of the environmental problems but would also involve people directly by engaging them on plantation or afforestation work and thus creating employment opportunities. By taking this programme on mass scale, the people can be made aware of the environmental problems and can be made self suffficient for their needs by including them to take up the planting in their own lands.

4.17 Decision of Government of India

The Prime Minister of India in his broadcast to the nation on January 5th 1985 said, "Continuing deforestation has brought us face to face with a major ecological and socio-economic crisis. The trend must be halted. I propose immediately to set up a national wastelands development board with the object of bringing 5 million hectares of land every year under fuelwood and fodder plantations, we shall develop a people's movement for afforestation." The Prime Minister's call for action and of making afforestation a people's movement at this juncture is indeed significant if ecological and socio-economic crisis is to be averted. However, such a call will mean a fresh lookmat policies, reformulation of new strategies, devising of new and more appropriate structures, and new initiatives that will make afforestation every one's concern and in ushering in a second green revolution which is what the greening of wastelands would be.

The urgent need for lacking such degraded lands which are to the extent of half the country's area was specifically addressed for the first time with the establishment of National Wasteland Development Board (NWDB) in 1985.

4.18 Objecives and Functions of tiie NWDB

The NWDB is supposed to deal with the following :

(*i*) Building landuse capability, specially at a district and village level aimed at integrated use and management of wastelands.

(*ii*) Conservation of selected ecologically agile areas like the Himalayas, the Western Ghats etc.

(*iii*) Regneration of degraded areas.

(*iv*) Grasslands and fodder development.

(*v*) Production and Conservation of fuelwood.

(*vi*) Promotion of agro-forestry, fauna-forestry.

(*vii*) People's participation at all stages and efforts to involve small and marginal farmers.

4.19 Schemes suggested by NWDB in The Eighth Plan

4.19.1 Technological Extension for Problem Lands

Identification of problem lands have been carried out in 146 districts of the country and wastelands maps have been prepared on the scale of 1 : 50,000. These maps have been distributed to the concerned States and districts. Detailed study will be conducted on small areas to cover various aspects of problem land *i.e.* comprehensive data base on physiography, hydrology, climatic and edaphic conditions, other ennvironmental and biotic factors operating on these lands.

4.19.2 Greening of Himalayas

The Himalayas which are covered with greeneries and supposed to be one of the largest tree cover hilly area in the world appears to have faced large scale denudation during the last four decades. This not only created Ecological imbalance but also extreme shortage of fuelwood for the local people and situation of major irrigation projects. Therefore the Eighth Plan aims at conclusion of the programme of reafforestation of the Himalayas on large scale, which would result in regaining the lost greeneries of the most beautiful largest hillock. This programme would be carried out in a phased manner with the selection of 10 to 12 districts at a time for intensive afforestation with the participation of the forest department, district administration, voluntary organisations and local villagers. The districts containing the catchment areas of major rasion systems, the rivers—Ravi, Chenab, Beas, Satlaj, Yamuna, Ganges and Sharda will be selected in the programme. The main object of this scheme is to maintain Ecological conservation of fragile watersheds and upper catchment areas of the Western Himalayas to improve the conditions of location and downstream and to increase the productivity of the local land and forest contributing to poverty alleviation in a large measure. In the national economy hill areas and plains are one organic system which are mutually dependent. Therefore ecological stability of the hills and sustainable life support system in the plains cannot be viewed in isolation. In this context active participation of the people becomes more relevant. To motivate the people extension programme will be followed in the Eighth Plan.

This scheme aims at conservation measures which include (*i*) more efficient use of firewood, (*ii*) increased grass production from grasslands and forest areas, (*iii*) effective fire control measures, (*iv*) substitution of wood for packing cases etc.

4.19.3. Restoration of Degraded Forests

It includes—(1) closure of the area to grazing and firewood collection, (2) closure and afforestation in case of degraded coniferous forests which do not coppice.

4.19.4. Afforestation on Rocky Areas

These are areas with severe constraints of climate and soil and afforestation is done with planting the species best suited to the site, by giving sufficient preparatory time for raising the grassland development programme should also be given weightage and involvement of the people is also needed.

4.19.5. Regeneration of Degraded Forest Areas

Degraded forest areas are very large in number and are continuously under biotic pressure to meet the subsistence needs of millions of poor living around them. In addition to this, there is a need of supplying the raw material to industry and seeking to serve certain ecological functions. Therefore, major programme needs to be launched regarding regeneration of degraded forest areas.

4.19.6. Social Forestry on Community Private Land

To increase the involvement of the people and to meet the fuel needs of the people, the main significant thrust of the Eighth Plan programmes for the wasteland development sector would be to start social forestry programme covering both public and private lands on which the production of fuelwood, fodder and low-value fruit plants will be done and raised.

4.19.7. Decentralised Planning for Integrated Use and Management of Wastelands

Implementation of Best plans and comprehensive programme needs a Best management at all levels and in all places in the country where this programme is being conducted.

4.19.8. Prevention of Wastelands Creation by Execution of Schemes by other Departments of the Government

Decelopment projects implemented by other departments in the Government like water resources, public works, transport, mines and heavy industry which are creating wastelands at an alarming pace should be consistent with the programme for reclamation of wastelands which is the direct responsibility of the Ministry of Environment and Forests.

The strategy of the Eighth Plan covers all issues regarding conservation and development of forests. It also stresses on satisfying needs of local people, Industry etc. The objectives and strategies of the plan are always comprehensive, but while achieving these objectives, the problems do arise in implementing the various schemes in the five year plan. But it is hoped that the expected results would definitely come out.

4.20 Proposed Outlay of NWDB in the Eighth Plan

The schemes which are proposed to be operated by NWDB and the proposed outlays are given in the following Table.

Table 4.1
SCHEMES TO BE OPERATED BY NATIONAL WASTELANDS DEVELOPMENTS BOARD IN THE EIGHTH PLAN

S.No.	Name of Schemes	Physical Targets Area(ha) million ha.	Crore seed-lings.	Total million ha.	Financial Outlay Unit Cost (Rs.)	(Rs. in Crores)
1	2	3	4	5	6	7
1.	Greeting of Himalayas	0.75	50	1.00	10000	1000
2.	Regeneration of Degraded Forests*	1.00	-	1.00	5000	500
3.	Reclamation of Problem Lands*	0.50	100	1.00	10000	1000
4.	Rural Fuelwood & Fooder Scheme+	0.25	50	0.50	7000	300

(Contd.)

1	2	3	4	5	6	7
5.	Integrated Wastelands Development Scheme	0.25	50	0.50	5500	275
6.	Decentralised Nurseries	-	300	1.50	2000	300
7.	Seed Development Agencies	-	-	-	-	25
8.	MFP-including Medicinal Plants	0.50	-	0.50	5000	250
9.	Aerial Seedling	0.25	-	0.25	1000	25
10.	Margin Money Assistance	0.25	50	0.50	5000	250
11.	Grants-in-Aid to Voluntary Agencies	0.10	20	0.20	6000	120
12.	Promotion of Community Structure for Wastelands Development*	-	-	-	-	15
13.	NWDB Headquarters	-	-	-	-	100
14.	TOTAL	3.83	620	6.95	-	4160

Note : 100% Central assistance to be given under all schemes.

* New schemes arising out of the restructured programme of NWDB being supported as pilot projects under Integrated Wastelands Development Scheme during 1989-90 and mergnng operation soilwatch.

\+ By merger of on-going centrally sponsored schemes of Rural Fuelwood Plantation, Silvipasture Farms and Area Oriented Fuel Fodder Projects.

The total coverage is established on the assumption of an average of 2000 plants per hectares in the case of the seedling production and distribution component.

4.21. Physical Targets and Proposed Financial Outlays

The physical targets in million ha. and proposed financial outlays from different sources under Social Forestry, afforestation and wasteland development are given below.

Table 4.2

PHYSICAL TARGETS (SOURCEWISE) IN MILLION HECTARE (P); FINANCIAL OUTLAYS (SOURCEWISE) IN CRORES OF RS. (F)

S.No.	Item of Works	NWDB Plans		RDD		State		Others		Total	
		P	F	P	F	P	F	P	F	P	F
1	2	3	4	5	6	7	8	9	10	11	12
1.	Reclamation of Special Problem Lands	1.0	1000	-	-	0.50	500	0.50	500	2.0	2000
2.	Greening of Himalayas	1.0	1000	-	-	-	-	-	-	1.0	1000
3.	Regeneration of Degraded Forest Areas	2.0	875	7.0	3500	1.0	500	-	-	10.0	4875
4.	Farm Forestry	2.45	645	1.0	240	7.55	1615	-	-	11.0	2500
5.	Community Plantations	0.50	500	-	-	0.50	500	-	-	1.0	1000
	TOTAL	6.95	4020	8.00	3740	9.55	3115	0.5	500	25.0	11375

CHAPTER V

PEOPLE'S PARTICIPATION IN COMMUNITY FORESTRY, FARM AND AGRO–FORESTRY

5.1 General

Afforestation annd greening of wastelands under different social forestry programmes has been recognised as one of the most important remedies against environmental degradation and the loss of common property resource. However, this programme cannot be successful without the active support of the people particularly in the rural areas. With more than 80% of the population in the rural areas, and more than half the area of India degraded, the need for the participation of support of the people is very much essential. The broad classification of the degraded land is as under :

(*i*) In forestry sector, out of 75 million ha. 40 million ha. are degraded out of which 30 million ha. are without tree cover.

(*ii*) In agricultural lands which are mostly privately owned, out of 143 million ha. about 40 to 60 million ha. are considered degraded.

(*iii*) The pasture and grazing lands and other public lands are most degraded as they did not get due attention for their restoration.

The distribution of the wastelands clearly indicates that the large proportion is under either the privately ownned agriculture lands or used by the rural commuity as a common or public land mostly managed by the village panchayats. The degraded forest areas are mainly around the villages which are solely dependent on these forests for the supply of fuelwood, fodder etc. All these factors clearly indicate that it is necessary to involve people in the treatment of the wastelands under private or community ownership. For raising plantations in the degraded forest areas around villages the cooperation of the people in the implementation of programme will have to be invited by creating some interest of the village community.

5.2 Community Forestry on Degraded Forest Lands, Community Lands and Village Common Lands

Traditionally the Forest Department was taking up the programme of afforesting of planting the degraded forests under their normal porgramme under working plans or under different Five Year Plans. However, there was hardly any involvement of the people. The people have no rights in the forest areas. At best they enjoy certain concessions which used to get curtailed whenever the department used to take up plantations. As the degradation of forests is more severe around the habitation, such areas are the priority areas. However, the results of the programme have shown that there was little success in the programme. This has led to the fresh thinking of involving people right from the stage of planning, execution, protection, harvest marketing etc.

The "Arabari Example" of establishing Forest Protection Committees for the protection of the Village forest having the root sotck and granting them some proportion of the income available from the periodic yield from the forest is the beginning of the involvement of the people in the actual management of the Forests at community level. More and more states are now thinking on these lines not for only protection but also for the afforestation and plantation under Community Forestry Programme. This programme would prove to be an effective means to provide substantial benefits to the village people by way of employment and other monetary benefits. It would also serve to check ecological degradation, and improving environment.

5.3 Experience of Community Forestry Abroad

The problem of scarcity of wood and wood products on one hand and the environmental degradations on the other hand is not only faced in India but it is a world wide experience particularly in the developing world. The Republc of China was the first country to embank on a major community forestry programme in 1950. Republic of Korea provide another example of a nation wide tree growing programme which achieved dramatic results.

5.4 Awareness at International Level

The farm and community forestry programme in developing world was also supported by a number of major international aid agencies. This is provided by the Publication in 1978 of the World Bank's forestry sector policy paper, with its acceptance of the need to promote forestry as a means of providing for the broader needs of the rural people. The World Forestry Congress held in Jakarta, also in 1978, added to the momentum being built up in the promotion of community forestry with its theme "Forestry for the people". The concept of community forestry was also actively developed and promoted by F.A.O. through its programme of "Forestry for the people". The scope objectives, and philosophy of this programme with its emphasis on self help and community engagement in small scale tree growing was set out in the publication "Forestry for Local Community Development" where it stated that :

"The objective is to raise the standard of living of the rural dweller to involve him in the decision making process which affect his very existence, and to transform him into a dynamic citizen capable of contributing to a wider range of activities than he was used to and of which he will be the direct beneficial—its ultimate objective is not physical but human. The physical goals which will be set are really means towards achieving the objective of enhancing the lives of human beings."

Above statement reveals the fact that community forestry and farm forestry are capable of providing employment. It will help in raising the poor's income by which they will be able to raise their standard of living and also be helpful in maintaining the ecosystem which is needed for human survival and sustenance.

5.5 General Psychology of the People

It is a matter of general experience that whenever there are large scale floods or drought in any region, society at once becomes aware of the bad effects of destruction of forestry and importance of the conservation and development of forests. But as soon as the severity of the calamity abates, the tempo slowly dies and soon it is forgotten. If we really have a long terms view of the contribution of the foestry sector quite a substantial amount spent on flood control, desilting, dams and canals, rehabilitation of flood affected persons and on scarcity works can be saved and utilised for other productive sectors. Adoption of the programme and of community forestry and farm forestry can be a long term view which will reduce the havock of the death of thousands of people and destruction of properties by calamities like large scale floods and drought etc.

5.6 Community Forestry—meaning and Need of Community Forestry

Community forestry programmes are based on the use of public or communal land for tree growing. Community forestry is becoming an accepted notation of land management in India and many other nations in the world. It is a needed response to the rapid determination of forest resources to the human poverty causing destruction, and to the untapped development potential of denuded uncultivated lands.

When India gained independence, its population was half the present size. Public lands were zoned for potential uses that appeared most beneficial at the time. Style of administration was different. Forest lands were viewed as governmental sources of timber and revenue. For transfer to private uses forest departments were custodians of the public reserve, regulating the release of forest beenfits more than managing forest lands for their developmental values. Revenue lands had nither public custodian nor manager. But now this trend has been changed and government, and people become aware with the important role of forests. They know that the traditional relationships of rural communities with forests and trees is largely based on a variety of goods and services they obtain from them, and also due to the fact that trees and forests form a basic part of their environment and their culture.

Degradation of the rural environments and consequent imbalance on fuelwood and other timber needs have been brought about due to the excessive population pressure on the ever declining resource base. Therefore, there is an immediate necessity to resotre the resource imbalance, by including tree cultivation as an integral part of a rural development programme. Therefore, Government introduces community forestry programme under social forestry scheme. The aim of the community forestry is to provide a forestry produce to the community as a whole. This can be achieved either by setting aside part of the areas around houses, fields, tanks, wells and along village roads, irrigation channels, railway lines etc., for growing trees, or by introducing multipurpose tree crops into the cropping pattern.

Community forestry has many attractions. In principles, they provide a means under which the landless class can associate themslves in tree raising and thus derive those benefits which otherwise are restricted for land owners. Certain programmes may be so designed that such benefits are deliberately chanelled towards the landless peasants. Community forestry is also one of the few approaches for tackling the problem of deforestation of village common lands brought about by the local people ; through mutual co-operation it becomes possible to protect these lands, bring about an increase in their productivity and also put an end to their gradual decline into barren wastelands. The most common form of community forestry programme is that in which the planning is done by the forest department. Other inputs such as seedlings, fertilizer etc., is also provided by the department. However, protection is the responsibility of the community.

The problem of wasteland is a serious threat to the nation and must be tackled on a war footing as a national problem. In order to preserve our produce land and forests and to develop wastelands, there is a crying need to create widespread awareness. Community forestry is one of the important instruments by which wasteland can be used for tree cover. Whenever, it is not possible to bring degraded government forest lands under green cover through the resources of the government, such land should be made available on long lease for afforestation to the grampanchayats, organisations of local advasis, educational institutes, NGOs (National Government Organisations) and even to private parties under specific conditions. In India now, under community forestry programme this approach is being adopted.

5.7 Operation of the Community Forestry

Community forestry remains a broad and fairly valued subject. It includes produce systems that may contain one or more timber species, forage and fodder crops, fruits, vegetables, grains, fibres, oilseeds or medicinal crops. It may cover uncultivated land described as private village, communal, revenue, reserve or protection forest. Community forestry is a generic term broader than forestry in the literal sense. Its identifying quality is that rights and responsibilities for land management are shared, in same degree, by the community and government and that governmental efforts are directed toward improving community welfare through better utilisation of degraded land. Planning and preparation for the community forestry is done in the same manner as in production forestry. In forestry, an assessment of forest ecology and productivity is made over a fairly large area. The estimates of annual growth is done from forest under the different systems, of the yields that can be sustained and the economic returns that can be expected under different rotations. Then plan of harvesting programame is needed which will maximise the sustainable economic returns, and develop a transport system to the extent that its costs are justified by the value of timber that can be brought to processing and marketing centres. Similar exercise is needed for the area under community forestry with only difference that the management unit is smaller.

Community forestry depends on both the capabilities of land and the choice of villagers about its use. The productive possibilities of uncultivated lands around villagers are determined by ecological conditions. Use of the area is determined by villagers' needs and bahaviour. Villagers may draw fuel from same parts of the area, bamboo and fodder of food from other. Some items may be for have use, other for cash sale. They may be drawn from lands under various tenures. In general, area for community forestry operations should be a circular area of village land use often near the village itself, very extensive at the ends of roads from the village centre, at the margins of the village area. The circular area can be expected to be relatively large, if the village population is large, the terrain is accessible, the land productivity is low and the village population is dependent on uncultivated lands for its livelihood. The circle can be expected to be smaller to the extent that the reverse conditions hold. This circular area around a village is the management unit for community forestry. The management unit includes the extent of uncultivated area influenced by village use. It is taken as such because all uses within it are interdependent. Changes of use in one part of the area will alter uses of the others. This concept is basic to community forestry, which is to reduce village use of forest lands.

5.8 Advantages and Limitations of the Programme

Communal programmes in principle, offer a number of advantages over individual tree growing. By using community lands and resources, they can permit landless people to share in the benefits of tree growing. They can also provide a focus for community action to halt the gradul degradation of comman lands through overgrazing and excessive wood cutting.

Many community programmes have run into severe problems. This is generally because they require a degree of commitment to a common effort which is often very hard to reach. But almost all the different types of community programmes offer opportunities for broad based socially and environmentally relevant rural development. The main lesson from recent experience is not that community programmes should be abandoned, but that their design and implementation should be imporved.

N.R.E.P./R.L.E.G.P. in social forestry scheme

N.R.E.P. is operating in all the districts of the country since October, 1980, with the main objective of generating additional gainful employment for the unemployed/underemployed in the rural areas by creating durable assets for developing the rural infrastructure. Social Forestry is one of the important activities under rural employment programmes. From 1981-82 onwards 10% of the annual allocations were earmarked for social forestry. This was raised to 20% from 1985-86, consequent upon the Prime Minister's broadcast to the nation on 5th January, 1985 wherein he gave a call to halt deforestation and the need to undertake a massive programme of fuelwood and fooder plantation with annual coverage of 5 million hectare and to develop a people's movement for afforestation. From 1986-87, 5% is required to be spent for promoting decentralised nurseries.

R.L.E.G.P. was launched in 1983-84 with the objective of improving/expanding employment opportunities for the rural landless so as to guarantee at least 100 days of work in a year for a family. The nroms of allocation of funds for social forestry are the same as that for N.R.E.P.

The physical coverage under N.R.E.P. has been 934698 hectares and total up-to-date expenditure Rs. 456.44 crores; similarly under R.L.E.G.P., an area of 452171 ha. has been covered at an expenditure of Rs. 201.55 crores.

Tree patta scheme was recommended in the conference of revenue ministers and revenue secretaries held in May 1985, developinng a people's movement for afforestation by creating people's interest through a grant of usufructuary rights in the tree planted. For this purpose a group was set up to evolve guidelines for a grant of tree patta/leases. The report of this group received in December, 1985, was sent to all the States in February, 1986, commending adoption with such modifications as may be necessary to suit the local conditions and for initiating some necessary action.

As a consequence of the tree patta scheme evolved by the Government of India a number of States, namely, Andhra Pradesh, Bihar, Gujarat, Maharashtra, Karnataka, U.P., M.P., Tamil Nadu and Rajasthan, have already started the scheme or a variant of it.

5.9 Funds Availability for Forestry Works

For implementation any development programme needs a lot of funds. Government made a provision of Rs. 2405.52 crores for afforestation activities during the seventh five year plan, the Rural Development Department alone contributing Rs. 1041.88 crores. R.D. funds, therefore, account for roughly 41% during the seventh five year plan. Recognising the intimate relationship between forestry and agriculture, the government has increased allocation for plantations from Rs. 1.28 crores in the first five year plan (1951-56) with a social forestry component of Rs. 16.41 lakhs to Rs. 452.20 crores during the sixth plan (1980-85). With a social forestry component of Rs. 351.88 crores, the programme aims to cover a total of 2.15 million hectares, of which 1.52 will be under social forestry, Rs. 62 crores under production forests. Thus not only has the total afforestation effort increased but the social forestry component has also increased in particular. In addition to these sixth plan programmes several state governments, particularly Tamil Nadu, Karnataka, Gujarat, Haryana and West Bengal have already started or are formulating new social forestry projects with the financial assistance of International agencies like the World Bank, Swedish International Development Authority (S.I.D.A), Canadian International Development Authority (C.I.D.A) and the U.S. Agency for International Development (U.S.A.I.D.).

5.10 Social and Economic Importance of Community Forestry

Half of the India, the poorer half, depends for its subsistence needs on the common lands, the uncultivated half of India. Over the years these lands, forest lands, grazing lands and other common lands have become highly degraded leading to deforestation and ecological degradation on the one hand, and poverty, unemployment and improverishiment and quality of life and druggery of women and children on the other hand. Unless this vicious circle of ecological degradations and poverty is broken, there is no hope either of achieving ecological balance or of improving the condition of the rural people. Community and farm forestry is the only answer to solve this problem. The plantation of trees under community and farm forestry plays a significant role. These trees dotted about the rural landscape around houses, along field boundaries and roadsides, and in communal grazing areas are of greater significance than the forest for the rural people. Their importance in supplying domestic fuelword is not negligible. In most developing countries virtually every rural family uses wood for all or part of its cooking and heating needs. India is not an exceptional case. If each and every family tries to plant 3 or 4 trees around their houses they will be able to satisfy their fuel need by their own efforts of wood products. In both rural and urban areas wooden building poles are used. Furnitures, packing cases, agricultural implementation and numerous other products are made by numerous small timber industries. These industries always use locally grown wood. Larger industries and such paper making rayon and match making industries also create a huge demand for locally grown wood. Animals and many communities depend on trees to provide foods, leaves and pods, fruit, nuts, mats, wall coverings and other woven goods. Considering all these uses and huge demand for wood there is a great need of increasing wood quantity by the way of community and farm forestry.

5.11 Farm Forestry and Agro-Forestry

Farmers as a community can take part in the social forestry programme by greening the community lands, forest areas, village common lands. However, individually also they can very well contribute towards restoring the ecological balance, and becoming self-sufficient for meeting their demands of fuelwood, fodder, small timber and requirement of wood for agricultural implements. As about 40 to 60 million ha. of private agricultural land is degraded, the individual farmer can take up tree planting in such degraded areas which are otherwise not suitable for agriculture. Sometimes the growing of trees can be combined with the agriculture by suitably adjusting the spacing of the trees to be planted. This can be called as Agro-forestry.

Farm forestry is the term usually applied to programmes which aim to encourage commercial tree growing by individual farmers on their private land. In these programmes, trees are regarded as a cash crop, and farmers are provided with assistance in growing them. This may include technical help, tree or subsidised seedlings, loan and various market support measures.

Agro-forestry has been defined as a sustainable land management system which increases the overall yield of the land, combines the production of crops (including tree crops) and forest plants and/or animals, simultaneously or sequentially on the same unit of land, cultural practices of the local population. Agro-forestry system is not confined to a particular region nor depends upon climatic conditions. It can be applicable in a wide range. However, recognition of the importance of agro-forestry as a production system has increased in recent years especially as regards its potential for optimising land use in the developing countries of the tropics. Agriculture is practiced for reasons of subsistence or for economic profit. The existing framework of policies and institutional infrastructure in the country attempts to make agriculture an economically convenient proposition. Agriculture sector can play a dual role by agro-forestry scheme. By cultivating an agicultural crop the farmer provides foodgrains to the country and by growing forest species he can earn much by making ropes, mats, wall coverings and other woven goods. Considering all these uses and huge demand for wood there is great need of increasing wood quantity by the way of community and farm forestry.

5.12 Profitability of Tree Growing

There is no doubt that under the appropriate conditions, where there is a strong market demand for wood, tree growing can be a profitable activity and many successful programmes carried out in India. The achievements in India have been particularly impressive. In a number of States demand for seedlings has expanded rapidly as a result of the high rates of return that can be obtained from commercial tree growing. In Gujarat, alone, nearly 200 million seedlings were distributed during the 1983 planting season.

A number of programmes have been based on encouraging tree growing by individual farmers to meet their own family uses. This is done in areas where commercial tree growing is limited by the lack of an adequate market demand of wood. Therefore, programmes of agro-forestry depend heavily on the careful identification of the tree species most suited to meet local needs and preferences. Their greatest potential almost certainly lies in promoting the cultivation of trees which provide multiple benefits rather than simply fuelwood.

5.13 Examples of Farm and Agro-Forestry

Farms of traditional agro-forestry is found in the home gardens of South-East Asia. These are characterised by a multi-layered mixture of a large number of food, fodder and timber species grown is in close association and yielding a wide variety of products. They commonly exist in small plots, located around individual dwellings. They are carefully tended and are often used for keeping poultry and small animals. Over the whole of Indonesia, tree gardening in various farms has been estimated to extend over more than a million hecatres (Wiersum, 1981). Home gardens play a particularly important role on the island of Java. Home gardens now cover 20% of the total land on the island. (page No. 31, Farms community forestry.)

In India these types of home gardens are not in existence. But some farmers who have farm houses and a plenty of land can preapre such types of home gardens around their dwellings plot on an experimental basis.

In East African highlands tall timber species are being planted and under the shade of these timber trees, coffee is grown in combination with banana or with beans. In same hilly areas of our country, some plots of forest land are transferred to tribals for agriculture. As said earlier, in hilly areas, the soil is generally not suitable to grow a foodgrain and it is uneconomical to cultivate only agriculture crop on the soil. In such lands agriculture crop with timber or coffee can be profitable to tribals. It will also be helpful in increasing the income of the tribals and raising their standard of living.

The ecological and economic advantages of combining trees with crops are numerous. In spite of these benefits, traditional agro-forestry systems are, nevertheless, coming under pressure in a number of parts of the world.

Acacia nilotoca is grown for fodder in parts of India. Pods of this species provide the major nutritional element and therefore they are collected by hand for feeding to animals. The traditional knowledge of rural people about the properties and uses of trees should not be overlooked by planners.

5.14 Trees as Cash Crops

In India, in many areas trees are cultivated as cash crops. They yield a variety of products including fruits, nuts, oils, species, fibers, medicines, and gums. In some States, rubber, coconuts, oil palm, and other tree based industries are a major element in the local economy. Cultivation of these tree crops is not confined to large commercial plantations, it also provides cash incomes and a means of livelihood for considerable numbers of small holders and marginal farmers. Babul tree is a typical example of a tree of multipurpose use.

The crops also provide important by-products. In coastal areas of India Coconuts grow in plenty. In addition to the production of copra for sale, cocunut palms yield a wide range of other locally useful products. The fibre from the husk, known as coir, is used in numerous small industries for making matting, rope, sacks and other products. Leaves are used for thatching, and the shells, husks, and leaf systems often supply a substantial proportion of local domestic fuel needs in sea coast areas, like Andaman and Nikobar and Kerala. Rubber trees are an important cash crop. They have a working life of about 15 years after which their productivity declines. As this stage, following a few years rapid tapping, called slaughter tapping, they are cut and used for fuel. Such wood from rubber plantations is the key source of wood fuel for towns.

In hilly area where coffee plants are grown with good grains, the prunings from coffee plants are also used as firewood in many areas and fruit areas can also provide fuel through regular pruning and firming, without affecting their productivity.

In dry regions some trees like Gum arabic tree (Acacia senegal) a cash crop tree, provide a multiple things to a community. It is a small leguminous tree. Most trees are produced from naturally established seedlings. It can be allowed to grow together with food crops and it can cut low and allowed to grow back during the fallow period, thereby helping to restore soil fertility. The gum is tapped during the dry season by cutting and peeling a piece of fark from the branches. It is used in food and textile production, medicines, and in the preparation of paints and printing ink. It has been traded for more than 4000 years. Sudan produces 80% of the total world output. In 1978, gum arabic accounted for about 7% of Sudan's total export earnings (G.I.T.E.C. 1979).

Acacia senegal not only provides gum but also a number of other locally important products. The foliage and pods are rch in protein and are an important feed source for camels, sheep and goats, particularly during the rainy season and early dry season. The pods are dried and eaten as vegetables. The hardwood is used for poles and agricultural implements, as well as for firewood and charcoal. A strong fiber can also be obtained from the trees' long flexible surface roots. These are used for rope and fishnets, and for living wells (National Academy of Sciences, 1980). In dry seasons, it will be profitable for Indian farmers to cultivate Acacia senegal on large scale. The local economy will also get boost by this cultivation and to some extent fuel problem can be reduced. In Sudan in recent years, the system of Acacia senegal cultivation is reportedly being abandoned in some areas. This is the result of the combined effects of shortening fallow periods and increased cutting for sale as fuel wood. (Hammer 1982). In other areas Acacia Senegal is being displaced due to the increased cultivation of groundnuts, wheat and other crops, using modern intensive agriculture techniques which provide a better cash return than the gum arabic (G.I.T.E.C. 1979).

5.15 Producing Wood for the Market

Farmers grow trees to produce wood for the market in some countries. India is one of them ; usually this is carried out on a relatively small scale and is normally centered on the production of poles, building timber and other high value wood products rather than fuelwood. For example in Maharashtra in Vidarbha region some rich farmers have started to take plantation on a large scale in their fields as a form of long term insurance. These are cut and sold for timber when money is needed for a wedding, a major purchase or some large cash outlay. Small farmers are also doing the same business by planting alternative species of timber such as mango trees in their farms. In Kerala, land holdings are generally very small and often consist of little more than a strip of garden around the house. Nevertheless, a large number of trees are grown. The principal species are cash crops such as coconuts and cocoa. Farmers often include a few timber trees as well. Even slow maturing species such as teak and Mahogany are sometimes grown as long term investment. In this State more popular species is ailanthos malabarica. This is in considerable demand from the match making industry and can be harvested after 8-10 years. The match industry in Southern India has been largely based on small holder tree growing for many years and some companies distribute tree seedlings to ensure a steady supply of wood. In Madras private plantations of casuarina are the major source of both firewood and poles. To meet the high demand for fuelwood which is scarce, local farmers started to plant casuarina for the market. Similar casuarina plantations exist around Bangalore, in Karnataka State, mostly within a radius of 60 km. from the city. They appear to be entirely in response to the demands of the urban fuelwood market. The report calculates the total fuelwood consumption of the city to be about 100 tonnes per day, of which an estimated 85% comes from private forests (Reddy 1983).

The above discussion reveals the fact that farm forestry or agro-forestry is an important source to overcome the various difficulties regarding wood scarcity.

CHAPTER VI

URBAN FORESTRY

6.1 Concept

The new Industrial revolution changed the economic pattern and living pattern of people also. In the first decade, just after independence, large number of industries were concentrated in few big cities. With the development of industrial sector in big cities or in urban area, people from rural area are being migrated to urban area in search of employment. As a result of all this industrial development in urban sector, the problem of pollution, and environment arises. All the government forests are wiped out and the environmental problem becomes more complicated. To solve this problem the concept of urban forestry emerged. Although the term "Urban forestry" is relatively new, certain practices and disciplines that make up the field of urban forestry have long been established. The use of trees to enhance the environment and many of the principles related to their use are extremely old. Trees have been aesthetically important to people since earliest civilisation. This concept was first introducuced in the year 1970.

Jorgensen states "Urban forestry as developed in Canada does not deal entirely with city trees or with single tree management but rather with tree management in the entire area influenced by and utilized by the urban population. This area naturally includes the watershed areas, the recreational area serving the urban population, as well as the areas lying between these service areas and politically designated urban areas and its trees. The politically established foundries for municipalities rarely include the entire geographical area influenced by urbanization."

In 1972, the Society of American Foresters initiated an urban forestry working group and put forth the following policy statement.

"Urban forestry is a specialised branch of forestry that has as its objective the cultivation and management of trees for their present and potential contribution to the physiological, sociological and economic well being of Urban society. Inherent in this function is a comprehensive programme designed to educate the urban populace on the role of trees and related plants in the urban environment. In its broadest sense, urban forestry embraces a multimanagerial system that includes municipal watersheds, wildlife, habitats, outdoor recreation opportunities, landscape design, recycling of municipal wastes, trees care in general and the future production of wood fiber as raw material."

The above statements clear the fact that urban forestry includes all woody vegetation within the environs of all populated places. A small town on the great plains is also an urban forest from the tiniest villages to largest cities. This means, urban forestry is not limited to only city places, but also to those lands on which trees contributed to the environment of populated places, such as roadside plantations, recreation sites, greenbelts in industrial area and contiguous suburbs etc. Any heavily populated area in the nation has elements of the urban forest. However, a small town of the great plains—an oasis of trees surrounded by open praire can also be considered an urban forest. One distinguishing feature between urban and rural forestry is that Urban forests and trees have more aesthetic value than rural trees. The distribution of urban forests in a particular geographic area is not difficult to determine. It is only needed to look at population centers in a city map to determine the larger part of the urban forest in and near the larger urban cities.

6.2 Need for Proper Landuse Planning in Urban areas

Within city environs it is necessary to look at the various land-use areas on which trees are likely to occur. Landuse pattern can be classified in residential use, park and recreation areas, transportation areas, particularly streets, agricultural area, institutional land, undeveloped areas and commercial and industrial areas where trees are often less abundant. Roads, streams, lakes, meadows and other open area (which is not under forest departments) should also be included in landuse pattern.

This landuse classification can also be made according to private ownership and public or private responsibility. Public lands include parks, street rights of way, highway and railroad rights of way, public buildings and grounds, extraterritorial lands, and riparian areas. Private lands include residential, commercial and industrial lands.

6.3 Practice of Urban Forestry Programme on Different Lands

6.3.1 Public Lands

Various projects can be implemented on public lands which are stated as follows:

6.3.1.1 *Urban Parks*

In more populated cities and town recreational parks should be set up to provide a source of recreation to the city dwellers for creating awareness giving information and knowledge about plant life and also keep the city air clean and free from pollution by serving as a green zone. In modern industrial era due to busy life people are always overstrained, relaxation is needed to reduce the strain. Only public city parks are not enough. The parks in remote area is a good recreation. These parks can be set up in the form of wildlife parks. With the help of these parts duel purpose can be fulfilled *i.e.* recreation and propagating the concept of preservation and protection of wildlife, through various ways. Developed countries have already taken steps in setting city parks and wildlife parks. This is one of the most revenue earning assets. In India there are national wildlife parks, but they are very few now, Government has taken various schemes to set up new wildlife parks and to modify and extent the available national parks.

These are the stimulant to promote the people for involvement in presserving and conserving the renewable natural resource for the common man. This is the duty of the present generation to preserve the rich heritage of wildlife. Urban camp sites are also one of the most important part of urban forestry. All modern well planned big cities have a special area which is kept aside for setting the urban camp sites having various plantations with the tree species of ornamental and asthetic value. Lawns specially planted with good grass are raised in these areas. These campsites are useful to provide a camping ground for tourists. It serves as a recreational site for city dwellers and also as a green helt. These types of campsites are created in western developed countries. But up till now India is far away from this creation. In India, where 60% people are still below poverty line, the priorities are also different. Though to balance the ecosystem these campsites are necessary but in India it may be considered as a luxurious item. Marginal productivity of a rupee incurred on production of fuel is much more than the marginal productivity of rupee incurred on campsites. But this does not mean that campsites, parks and wildlife parks should not be set. It will take its own time for this type of development. India is a developing country and has to achieve a developed stage in the early future. This is the reason, these type of campsites are very few in India. New Delhi has taken the lead in developing such forms of urban campsites.

The another part of Urban forestry is wilderness trails. This idea was developed for the first time in U.S.A. Wilderness trails are forest areas which are set aside exclusively for conservation and recration purposes. Such areas are usually selected in the vicinity of Urban centers, so that the city people may visit them and derive maximum benefit. Street rights of way constitute a substantial part of public urban forest lands. These are strips adjacent to streets or medians between divided boundaries. They can be in the form of trees, lawns, parkways, or parking strips. Street side rights of way vary in width and often provide space for side walks. Street side trees planting programme is implemented and administered by city forestry departments. This programme can also be implemented by municipal corporation. Implementation of programme is not important but the most important factor is the maintenance of planting trees. There is always enthusiasm in the beginning (for planting trees) but maintenance is always neglected because of the scarce funds.

6.3.1.2 *Highways and Railroad Rights of Way*

In India every city and town are connected with highways and railroads. The land adjacent to these arteries, at intersections and in medians can often support a significant part of Urban forestry. There is no significant progress in planting trees and also maintaining the plantations already made in these areas.

6.3.1.3 *Public Building and Grounds*

Ground adjacent to public buildings such as schools and colleges, hospitals, and auditoriums, museums, penal institutions, courthouses and other Government office buildings, military installations, cemeteries, airports, guidances are important parts of urban forestry. In India in these buildings only gardens are developed, but there is no plantation of forest trees species. In America, grounds in these buildings are well landscaped with generous plantings of trees and shrubs. These segments of the urban forest can be owned and managed by all levels of government and represented by a myriad of departments and agencies. There should be an Interlinkage amongst all Government departments for the development of urban forestry.

6.3.1.4 *Extraterritorial Lands*

Land adjacent to the cities and town, owned by public authorities can be covered with forest tree species in the form of green belts, graves, forest preserves and even vegetable land fills. These forests are extremely valuable to the urban environment as they may provide watershed protection, recreation, scenery solace or a place for disposal of waste products. In India many big cities and towns which are over-populated and also industrially over-crowded have wasteland beyond the boundaries of these cities and towns. Under social forestry programme this land can be covered under this scheme.

6.3.1.5 *Riparian Area*

River fronts, canals, levels, channels, diversions, lakesshores and even seashors are parts of the urban forest. These lands are developed as parks or recreation areas. But considering the scarcity of fuelwood, these lands can be used for fuel species along with recreation-parks. Such water related areas are extremely varied, and trees are important elements when their presence is not inconsistent with the primary use of the land.

In India where 60% people do not use domestic gas and are dependent on fuel trees species, recreation motive is not important but they should be supplied this type of fuel and therefore in urban forestry programme along with recreation motive all public waste-land can be developed with plantations for mixed purpose *i.e.* parks with fuel species.

6.3.2 Private Lands

All kinds of residential, commercial and industrial lands can contribute to urban forestry. Investments made by private land owners on these lands, may fulfil the both objects *i.e.* fuel supply and recreation also. Residential areas vary from highrise apartment complexes to single family dwellings, and tree species associated with both types of areas vary according to appealness of house owners, residential areas constitute a substantial part of urban land use.

Commercial areas make up a less significant part of the urban forest. Industrial areas can often be more utilitarian than beautiful. However, industrial areas are capable of producing fuel in sufficient quantity in the unoccupied areas. The green belt made there world also keep the pollution in check.

6.4 Environment of Urban Forest

The urban forest exists in an environment dominated by people. Thus the forests must grow in the spaces left after all the structures necessary to urban society have been built. The forest must be moulded to fit these spaces as it cannot interfere unduly with streets, side walks, parking lots, wires, pipes and other objects. The forest must also withstand compacted soils and polluted air, and it must be compatible with people, pets and vehicles. The environment of the urban forest, then involves site, space and people.

The site environment involves the location of trees. Most of the time in absence of plain area the cities have grown by removing the adjacent forest. The sudden removal of strips and patches of a natural forest creates edge at which previously shaded trees are exposed to open sunlight and wind. Surface of the soil drainage is also damaged. The trees to be retained are often damaged by physical contact with machinery by root exposure or coverage, and by indiscriminate pruning. In development process of cities the forest is considered as liability and some developers have considered it to be removed and the land is levelled.

6.5 Need for Proper Planning of Urban Forestry

In urban area space is the the most critical factor. The big cities or towns are always over crowded and space is limited physically by buildings, neighbouring trees, overhead wires, curbs, sidewalks and underground utilities. It is also limited by other spaces appropriated for line of right vision for traffic signals signs, street lights and for parking sites.

Therefore planning of cities should be such that road side plantations, big parks with flowering plants, also some fuel trees to provide people's fuel need should be given preference. Urban trees should possess a suitable form for its available space, and it should have roof system compatible with soil space restricted because of concreate and underground utilities. The combination of small trees less than 30 feet (9.14 m) at the time of matarity, medium trees 30 feet to 60 feet and large trees 60 feet over should be adopted for planting purposes. The trees which are narrow at the top have poor shade. Such type of trees are not useful for roadside plantations. These are useful from the aesthentic point of view. The texture of the soil, water tables are often changed, because of the alternations in construction activities. Some additions in the city planning, special soil treatments and ammendments are often needed to assure plant growth. Soils can be physically improved by incorporation of sand or replacement of new soil etc. Soil acidity must also often be altered as it largely governs the availability of nutrients.

The microclimate of the cities, pollution and air pollution of the cities influence the trees growth. All these factors are influenced by human structure. Physicial, chemical, and foreign substances, transportation, dlectricity generation and industry are the primary sources of air pollutants. Pollutation damages the tree growth. But once the trees are mature they can help to reduce pollution. It is the beginning; adverse effect of pollution on trees growth is minimised. The later effect of trees on pollution will be beneficial to human society.

6.6 Benefits of Urban Forestry

The urban forestry is important to the city dweller in many ways. Trees provide shade, beauty and a long list of other benefits.

There can be four broad categories of various benefits *i.e.* Climate Ameliorization, Engineering uses, Architectural uses and Aesthetic uses.

Solar radiation, air temperature, air movement and humidity are major elements which are associated with human comforts of these four elements. New inventions in Engineering faculty always control the compact zones created by human efforts. Engineering aspects is an important aspect which affects the urban forestry benefits. We control the comfort zone very precisely in buildings. With a twist of a knob or push of a button, we can regulate indoor temperature, light, humidity and air movement. If similar controls could be applied to the out of doors environment then it would be possible to regulate a mocroclimate to a degree by the proper use of trees and shrubs and it may ameliorate the climate sufficiently for us to be comfortable.

6.7 Composition of Urban Forests

Urban forestry is in infant stage in India. In India this concept of urban forestry has not been developed Therefore, it is impossible to discuss the total composition of urban forest. This urban forest can be developed in both ways, *i.e.* natural and manmade. Some specific areas in like hill stations such as Nainital hilly forest in U.P., Oti's forest in Karnataka, Matheran in Maharashtra are the examples of natural forest But most of the part of urban forestry in big cities and town is a manmade forestry. Hill stations are always used for recreation. The efforts are being made to develop all hill stations in India for recreation. Therefore the forest in hill stations can be considered as urban forest which is natural forest.

The urban forest includes also a large area of native woodlands, both within and adjacent to cities and towns. This forest is manmade forest. In big cities like Delhi, Bangalore, Nagpur, Pune, we find the recreation parks and roadside plantations. Green belts in cantonment areas and also in municipal areas like Chakarata in U.P can be included in Urban forest. There are large areas covering shrubs and small plants in every city and towns which can be converted into beautiful forest area in the form of urban forest.

While developing urban forestry the uses of plants are more important. The use of specific species is always influenced by soil composition, climatic differences and the availability of water resources. Native woodlands in cities, and towns with indigenous species can be developed in the form of urban forest. In U.S.A. in many cases, native forests have been urbanised as cities have grown into them. In these situations the original forests have been so altered by the supplemented planting of introduced species that they have taken on the characteristics of manmade forests. In India as due to the increase in population growth and the establishment of various industries, the cities have grown and crossed their limits which resulted into forest boundaries, so it will not be injustice in reconsidering this adjacent forests as urban forest.

Trees, shrubs and grass ameliorate air temperatures in urban environments by controlling solar radiation. Tree leaves intercept, reflect, absorb and transmit solar radiation. Trees and other vegetation also aid in ameliorating summer air temperatures, through evapotransipration. Trees have been called nature's air conditioners. A single isolated trees may transpire approximately 88 gallans (400 liters) of water per day. This has been campared to five average room air conditioners. At night heat is lost primarily due to the exchange of infrared radiation between city surfaces and the atmosphere. Tree canopies slow the loss of heat from city surfaces, providing a screen between the cooler night air and the warm surface materials. Thus, the night temperatures are higher under trees than in the open (Federer, 1976). In urban areas, this differential may often be as great as 10 to 15°F (5 to 8°C) (Federer, 1970). Deciduous trees intercept the Sun's rays during the summer months and allow them to pass through during the winter.

Air movement or wind also affects human comfort. The effect may be either positive or negative depending largely on the presence or absence of urban vegetation. In India, most of the part of the country is too hot in summer and some big cities like Bombay, Calcutta, Madras surrounding seashore have a humid atmsophere. This hot temperature and humidity can be minimised by planting trees and shurbs in big cities. Trees and shrubs can help in increasing the cooling effect in hot areas in summer and as well as in decreasing the humidity of the seashore areas. Trees and shrubs control wind by obstruction, guidance, deflection and filtration ; effect and degree of control very with species size, shape, foliage, density and retention and the actual placement of the plants. Trees by themselves or in combination with other barriers may alter airflow over landscapes and around buildings. It can obstruct undersized as well as desired air movement through buildings. It depends upon the situation of the planting trees and shrubs around the roads and buildings.

6.8 Engineering Uses of Urban Forestry

Engineering uses include soil erosion control, air pollution, noise abatement, waste water management, traffic control and glare and reflection reduction.

Wind erosion and soil erosion are mostly connected with rural agricultural lands where large land surfaces may lie unprotected, where winds are frequent and strong and where precipitation is minimised and unpredictable. In urban forest, water erosion is more important than wind erosion. Soil erosion in urban areas is usually associated with construction activities where soil surfaces have been exposed. In absence of urban forest soft areas in the soil surface may be removed faster, resulting in rills and troughs. Plants reduce water caused soil erosion by intercepting rainfall by holding soil with their roots and by increasing water absorption through the incorporation of organic matter. In addition, plants are more attractive than mechanical water erosion control devices.

6.9 Urban Water Disposal and Urban Forestry

In India in big cities rapid population growth accompanied by an expanding industrial base has greatly increased the demands on our nation's water resources. This growth is also creating ever increasing water disposal problems. Reuse of waste water is possible, if it is purified and diluted ; some treatment is needed for purification and diluting process. In U.S.A. waste water disposal begins with primary treatment. "Large objects are removed and the remaining material is ground in a grit chamber and piped and suspended organic material are allowed to settle. After the water is skimmed, it is discharged into a lakes or streams. In primary treatment one-third of the bio-chemical oxygen still remains ; therefore, secondary treatment is also applied. This treatment is primarily biological in nature and involves the use of microorganisms to decompose organic compounds. Recently a third process called a tertiary treatment has been developed to aid in nutrient removal, renovation and reuse of waste water. This process is more widespread in the United States. These all treatments are very costly, but though in India there are such big plants of purifying the water, Availability of large blocks of land is needed for this process. Forest land and agriculture land can be used for disposal and forest preserves, green fields scenic parkways and perhaps even highway medians can be used for waste water disposal management. As per Forest Conservation Act 1988, green belt is compulsory for each and every factory and industry. These green belts can be available for above process. If recreation parks, road side plantations are made in urban areas, this land can be used for waste water disposal.

6.10 Noise Abatement and Urban Forestry

Noise is commonly thought of as excessive or unwanted sound. Every city is accompanied with full of noise. This noise is known as invisible pollution. It involves both physical as well as psychological effects. To nutralise this invisible pollution, the movements of sound can be restricted through vegetation over which the sound is passing. Vegetation serves as unnatural barrier to minimise a transmission of sound waves through the air.

How effectively plants control sound levels is determined by the sound itself. The planting involves the climatic conditions. The properties of sound include type, origin, decibel level and intensity,. Planting factors include the species, arrangement in relation to the sound source and the receiver and planting height and density. Climatic factors include wind direction, velocity, temperature and humidity. In addition to deflection, absorption, refraction, reflection and masking plants may also attenuate sound by their influence on local climate *i.e.* their stabilization of temperature, their modification of wind spread and other factors. Weiner and Keast (1959) pointed out that sound propagation patterns over woodland areas differ from those over open level terrain.

The study conducted by the School of Engineering at the University of Nefraska and the Rocky mountain fore and Range Experiment Station, U.S. Forest Service revealed the fact that the potential of trees and shrubs for sound reduction is quite promising.

To reduce the noise of the vehicle passenger, and a single row of dense shrubs backed by a row of taller trees are useful in urban residential areas. Noise from high speed traffic can also be minimised by planting several rows of trees and shrubs with trees of sufficient height adjucent to the traffic lanes.

As discussed above plants can play a role in noise abatement. Though there are limitations to their effectiveness it is impossible to eliminate sound completely, but it cannot be considered a wastage of expenditure incurred on tree planting in urban areas.

CHAPTER VII

FORESTRY : SOURCE OF GENERATION OF INCOME AND EMPLOYMENT

7.1 INCOME FROM FORESTRY

Forest is an important source of income and generation of employment also. Forest is a factory of wood production just as an Industrial factory. But there is a difference among Industrial goods factory and forest factory. Forest is an industry in which capital is invested by nature *i.e.* the society gets this capital at tree cost. Every year we get wood produce which is the interest of the tree capital which is also a tree gift of nature. Every year each tree is putting on increment in volume which can be considered as interest of a tree capital. But in Industrial goods factory the capital; is invested by man *i.e.* man-made capital and this capital is embodied in the produced goods which is invisible in the commodity. But tree capital is always visible in wood produced goods.

To obtain maximum income every industry must have an efficient management. In the same way forest industry also needs an efficient and good management to maximise the revenue with the implementation of conservation policy and not a commercial policy. According to William Duer, Forest Economist, "Conservation is nothing but saving trees and forest stands grow in value by accumulating growth from year to year." It is just like the interest on bank account. The people or firms put their savings in the bank by foregoing current cunsumptions in favour of future enjoyment and they get a reward in the form of interest on their capital. In the same manner the forest management envisage savings through conservation. This saving, in case of the growing stock of forest, is also nothing but investment. Therefore, conservation does not mean keeping the forest as they are without taking any Economical advantage, it means the judicious exploitation of the forest by proper management. In simple terms forest conservation can be defined as "growing and management forests so as to obtain maximum yield of timber, wildlife, water shed production and other values without deterioration of forestes".

7.2 Forest Management

Forest management has been defined as the practical application of scientific, technical and economical principles of forestry. The application of the scientific principles included the knowledge of silviculture which has been defined as the art and science of cultivating forest crops. The standing crop in the given area is called "growing stock" which is the main source from which different forest products are obtained. Since the management of forests initially started with production of wood in the form of timber, pulp wood or firewood and also bulk of the income from the forests is from wood products, Forest management mainly deals with timber production and the regulation of yield of wood from forest.

The management of any enterprise or undertaking involves policy organisation and conduct of all operations to fulfil the objectives of the owner. The forest management is no exception to this general principle of manaegment. Regarding management, policy matter is discussed in the national forest policy 1952, 1971 and 1988, which is discussed in detail in the chapter 3.

Most of the forests in India are owned by the Government. However, forestry being state subject the management rests with the Forests Departments of the State Governments. To have a uniformity in forest management, the Government of India declared National Policy in the year 1952. It proposed the classification of forests on a functional basis into (*i*) protection forests, (*ii*) national forests, (*iii*) village forests and (*iv*) tree lands and emphasised the need for evolving balanced and complementary land use. It also made provisions for ensuring progressively increasing supplies of grasing, timber for agriculture implements and firewood to rekease cattle dung for use as manure. The policy also discourages the indiscriminate extension of arable lands by the excision of forests as this not only deprives the local population from forest products like wood, grass etc. but also deprives the land of its natural defence against hot wind, dust storms and soil erosion. The policy prescribed that the national forests were to be managed on the principle of progressively increasing sustained yield to meet the requirement of defence, communication and industry. From the general principles laid down in the National Forest Policy, it will be clear that the policy emphasises the protective aspects of forestry and also the welfare of the local population by way of the services the forests can provide. It also aims at conservation forestry in which the object is to get maximum sustained yield from the forests.

7.3 Principles of Forest Management

The guiding principle of forest management has changed from time to time. Two major concepts which are accepted in India for forest management are (*i*) Sustained Yield and (*ii*) Progressive Yield.

7.3.1 Sustained Yield

The notion of sustained yield was developed by German Foresters. Since the foundation of Indian Forestry were laid by eminent German foresters like Nrandis, Ribbentrop and Schlich, it is but natural that in Indian forestry sustained yield became the focal point round which all planning is done. A sustained yield may be defined as the regular and continuous sully of the derived produce to the full capacity of the forests. This definition clearly includes all forest products including major products (wood) and minor forest products. When it is applied to major produce, it implies that the removal from year to year should be approximately equal or progressively increasing to the extent feasible and in no case the removal should exceed the productive capacity or the increment of the forest. The trees put on a layer of wood on themselves annually and this annual growth of trees in the forests is the annual increment. The removal of this increment from the forests without the loss of original tree growth or capital would, therefore, actieve the object of sustained yield.

7.3.2 Progressive Yield

Slowly however, there has been a shift from the concept of "Sustained Yield" to "Progressive Yield". This concept envisages—

(*i*) raising productivity of soil

(*ii*) raising productivity of the stand by

(*a*) judicious tending,

(*b*) enrichment by changing the composition of the crop,

(*c*) replacement of original interior forest into a valuable forest largely through artificial regeneration.

This major change of policy from conservative "Sustained Yield" to dynamic programme of "Progressive Yield" is reflected in the recommendations of the National Commission on Agriculture. The Commission in its Interim Report submitted in 1972 on Production Forestry and Man-made Forests recommends:

"There should be change over from the present conservation oriented forestry to a more dynamic programme of production oriented forestry. Considering the advantage of aggressive man-made

(*a*) *Long Period of Production*. Timber production is a long term process involving a period about 50 to 150 years. Therefore, it involves uncertainties and risks regarding price, technology and cost.

(*b*) *Duel Nature of Product.* Though each tree is putting on increment in volume it is impossible to peel off each tree's layer of new wood each year. The whole tree has to be cut to harvest any product. But tree or the growing stock is a unit of timber producing factory. Harvesting it, therefore, liquidated part of the factory itself. Thus timber product is also a timber growing factory and vice-versa.

(*c*) *High Ratio of Inventory to Output*. The ratio of timber production to annual growth may be as high as 1:50. This feature of timber leads to high degree of flexibility in harvesting and marking. In agriculture, the farmer has no alternative but to harvest the entire crop in a given season. However, in case of forestry the forest manager may increase the harvest in a year by drawing stored inventory or decide not making any harvest in which case the growth accumulates in the capital and the production in a particular year can very according to the decisions taken regarding harvesting.

(*d*) *Joint Production.* 'To grow timber' may also involve production of pulp wood, firewood and may influence the quality and quantity of run off, water regime and other indirect services.

(*e*) *Low Value Productivity.* The rate of return from the land used for timber production is less than the market rate of interest.

These peculiarities have to be kept in mind when planning for the execution of different works involved in forest management.

7.4 Forest Management Plan or Working Plan

Forest Management plan or working plan is written scheme of management of the forests aiming a continuity of policy and action and controlling treatment of forests. While the strategic planning involves the formulation of policy and objectives of planning, the working plan gives the details of technical planning, involves in achieving the general objectives. Generally a trained forest officer is assigned a job of preparation of the working plan. The working plan officer collects maps of the forest area and inspects the forest through his assistants. The details of the forests regarding species composition, density of the crop, approximate age of the crop etc. are mapped. The working plan is based on the type of the field data obtained from making inventory of trees or enumeration of trees, determining the growth rate of important species, classification and mapping of forests and finding out the area under each of them. If the forest area in question had already been under some working plan which is to be revised, the working of the previous plan is critically analysed and evaluated. The plan discusses the utilisation pattern of the produce and assesses the demand for different products. The plan also includes the logging methods, past and current rates and prices and such other data which can serve as a guide for some years into the future execution of forestry works. Based on the facts collected by the working plan officer, the plan is prepared for a given period of time.

A well prepared plan indicates which area of forests, species, their number and sizes of trees may be scheduled for cutting in years ahead, where planting is necessary, what areas of forests need improvement in stand and what forests need to be kept standing for protection of the site conditions. After a plan is prepared, it is scrutinised at different levels of administration before it is finally sanctioned by the Government. Generally such document is prepared for a period of 10-15 years. However, the revision of the plan requires lot of time and, therefore, provisions for the period beyond the period of the plan is also made.

Certain forest areas which were not under scientific management are also vested in the Government. For such type of areas a " Working Scheme" is prepared. It is a document similar to working plan but the plan of management is more detailed.

7.5 Yield Regulation

Every working plan or working scheme gives the method of yield determination or calculation of the amount of material which may be removed from the forest annually or periodically. It also gives the annual area over which fellings are to be made consistent with the objectives and the silvicultural system adopted. The regulation of yield ensure acting of the trees at maturity and to secure maximum yield possible every year or periodically without the forest being cut or overexploited endangering the future yield. The regulation also aims at limiting the area to be cut annually to that which can be regenerated either naturally or artificially.

To fell yearly the exact quantity specified by yield regulation in working plan is not practically possible.

Unforeseen contingencies, silvicultural or economic, often necessiate felling of more trees in one year or less in another year. Therefore, a sustained yield over a period rather than each year is an attainable proposition and is generally followed in the management plans or schemes.

The above discussion clears the fact that conservation approach is the most beneficial from the long term view.

7.6 Classification of Forest Economy

Forests contribute different products and services which may include several thousand items. For simplicity in analysis it is possible to classify them into a relatively small number of homogeneous groups. One way is to classify the benefits into "direct" and "indirect" benefits or contribution of the forest. Another most widely used method based on economics is grouping the products and services into two main classes of primary and secondary benefits. The primary benefits are dregarded as those which are restricted to the direct recipient of the service while the secondary benefits are spread over the whole economy of a very large sector of Economy.

According to Exchange characteristics forest products and services can be classified into four categories.

(*i*) Market determined goods.

(*ii*) Non-market determined goods.

(*iii*) External economics.

(*iv*) Intangibles or social and political services.

From these four categories the visible benefits from market determined goods can be measured as benefits from the remaining in terms of money only. Categories cannot be measured in terms of money. Therefore the income received form market table goods is considered. Invisible benefits are discussed in the Environment Chapter.

The market determined goods include all the commodities which are mentioned as direct products and which are terms as "Forest products". All these products are exchanged through a market system. Because their values are market determined. The contribution of forestry and forest products to the national income can be readily measured in monetary terms. Generally the share of national income creditied to the forestry sector comprise the value of the forest product.

Forest produce includes wood production, minor forest produce and plantation activities.

Table 7.1

NET DOMESTIC PRODUCT (AT CURRENT PRICE)

(1960-61 to 1978-79)

(Rs. in crores)

Sector	1960-61	1965-66	1970-71	1975-76	1976-77	1977-78	1978-79
Agriculture	6580 (49.3)	9534 (45.8)	16354 (47.4)	25808 (41.8)	26692 (39.8)	30359 (40.4)	31023 (38.6)
Forestry and Logging	174 (1.3)	317 (1.5)	397 (1.1)	722 (1.2)	830 (1.2)	935 (1.3)	1122 (1.4)
Total (Net Domestic Product at Factory Cost)	13335 (100)	20801 (100)	34519 (100)	61899 (100)	67120 (100)	75058 (100)	80090 (100)

7.7 Income from Forests, Wood Production

Wood production is the main source of income from forests. The Production of industrial and fuelwood is an important indicator of the forest development in the country.

7.8 Minor Forest Produce

Apart from the timber and fuel forests are supplying a variety of products which are grouped as minor forest produce. They include Bamboos, Canes, tendu leaves (used for manufacturing bidis) grasses, essential oils, gums, resins, medical plants, fatty Lac, tanning material, dyes, animal hides and horns etc. Quite a few of them are foreign exchange earners.

7.9 Plantation Activities

Plantation and afforestation programmes under man-made forestry form very important component of the national programmes for renewing the forest resources. These programmes lay special emphasis on planting fast growing commercially important species like Eucalyptus, Teak, Bamboo etc. and other species also like. which are being planted under social forestry programme. There are various afforestation and plantation schemes under state plans. By undertaking plantation schemes dual objectives are being achieved *i.e.* conservation and growth of forest and income generation.

Table 7.2

REVENUE PER HECTARE FROM FOREST AREA OF SELECTED COUNTRIES

Country	Gross Revenue per hectare (in Rs.)	Net Revenue per hectare (in Rs.)
India	21.50	11.50
Austria	336.00	80.00
United Kingdom	NA	140.00
West Germany	565.00	130.00
Switzerland	494.00	190.00

Note : NA—Not Available.

7.10 Employment from Forestry Sector

As discussed earlier unemployment is a severe problem which is not being solved though, lot of job opportunities are being created every year. Forestry sector is one of the most important sector which provides alternative employment particularly in the rural areas.

India has been successful in increasing food production spectacularly in the last twenty years. Yet even today nearly half the population go hungry not because of insufficiency of food produced but because they cannot afford to buy enough. They do not have money because they are unemployed or underemployed. There is an alarming increase in the population of India. Therefore, not only now but also in future, unemployment will constitute a very serious social and economic problem.

The problems of low labour utilisation and low earning particularly in the backward rural areas have become the central preoccupation of the economists, planners and the Governments. The current and prospective growth in labour force which is really massive has led towards employment oriented economic development programmes. Thus, among the objectives of the present economic and industrial policies that tend to attract attention are the twin goals of employment creation and through it the improvement in the income distribution especially amongst the less fortunate sectors of the society.

7.11 Role of Forestry in Social and Economic Development

To analyse and discuss the role of forestry and to identify the potential contribution of forestry to social and economic development of Asia and Far East, a consultation on employment in Forestry was held at Chiangman, Thailand in 1974, under the joint auspices of F.A.O., I.L.O. and S.I.D.A. In the deliberations of the consultations it has been recognised that employment problems of the countries of the region were very serious and the situation was getting worse. Apart from a productive factor, employment is a channel through which the population derives benefits from the economic system. The only resource of the people without capital or land is to sell their capacity to do work. If this cannot be done, they are isolated from the distribution of wealth through country's economic system. Thus there are strong links between employment, poverty and more equitable distribution of income indicating that the creating of employment is in itself an important factor for social and economic development. Forestry is important not only as prominent source of employment generation but also as a vehicle for social and economic development in rural areas. Important recommendations, of the Thailand Consultations are give below :

(1) Recognition at all levels of planning of the potential of forestry to generate employment and its contribution to national development.

(2) Promotion of self-employment among the rural population in the collection of minor forest produce through organisational support and the development of market facilities.

7.12 Features of Employment in Forestry

The total output of any country depends upon the productivity of its labour force and on the number of people at work. The people constitute one of the principal resources in a developing country like India for expanding production. The capital being scarce in India, the obvious choice is the choice of "Labour Intensive" production process and other forms of economic activity. It has been recognised that the imbalance in creation of jobs and growth in industrial output in the developing countries is mainly due to the excessive use of capital intensive techniques of production. This has resulted into rapid urbanisation leading to urban unemployment.[3] Forestry sector is able to make some contribution to solve this problem, mainly because it has vast material resources and recurring operational activities which are mostly labour intensive and are capable of giving employment to many unemployed or underemployed persons. Forestry sector is highly labour intensive. The ratio of wages to material/capital may be as high as 9 : 1. It has been estimated that one job could be created with an investment of a meagre sum of Rs. 3000 only. It is possible to provide employment for many millions for productive work of forestry which does not involve the monotonous drudgery of breaking metal for road mending which the Government generally provides as a part of relief measure thought out by some district officer of British Raj hundred years ago.

In addition to the creation of large number of jobs in the operational activities in forest, other important features are :

1. *Unskilled Jobs.* Most of the jobs provided by the forests are semiskilled and require a little or no training.
2. *Suitable for all.* All types of workers including men, women, boys and girls can be employed on different forestry work involved in the management of forests.
3. *Dispersed Employment.* The employment provided by forestry sector is generally in the forest areas which are remote areas. Therefore, the employment provided is dispersed. With this dispersed employment the pressure on transport, housing, investment, etc. which are the main social problems of urbanisation are considerably reduced.
4. *Local Jobs.* The availability of jobs locally reduces the exodus of unemployed labour to urban work centres. It also helps to bring the monetary economy to the rural areas and so has an impact on distribution of income.
5. *Low Cost.* The real cost to the society of employing unemployed people is very low, often lower than the cost of the wages. Therefore, these people can be rationally employed in activities such as planting, road-building where wages account for a large part of the total cost.

7.13 Appraisal of Employment Potential by National Commission on Agriculture

While discussing the case for revision of the National Forest Policy, 1952 the National Commission on Agriculture has observed that the policy should clearly define interrelationships of forest economy with rural and tribal economy. The factors to be considered are (*a*) employment, (*b*) right of user, (*c*) involvement of local people. The involvement of the local people in forest development may be useful to contain the forces of destruction of the forests. Forestry has a great scope for absorption of rural labour. The multiplier effect of employment also works to the advantage of rural sector.

The Commission further observed that the employment offered in the forestry sector was often in off season for crop production and was, therefore, complementary to employment in farm production. For example, logging is generally carried out in lean agricultural months. As the forests are generally located in backward areas, the employment of people in forestry activities will directly benefit the backward communities like tribals and other rural agricultural labourers. According to the report of the National Commission on Agriculture, employment in 1970 is estimated in forestry sector to be 2.3 million manyears. It is estimated that in forestry sector as a whole, the additional rural employment generated will be of the order of 5.5 million manyears by 2000 A.D. In addition there is a vast scope for employment in rural areas in transport of forest produce of forest based industries such as saw-milling, pulp, paper panel products, matches, katha, resins etc.

7.14 Forestry as an Instrument of Social Change

The draft of the Sixth Five Year Plan has recognised the key word of development as employment and more significantly rural and tribal employment and the tribal employment has a key source in forests alone. Had we reforested an area of one million hactares as recommended by the Fifth Five Year Plan we could have created an employment of 250 million mandays during the Fifth Five Year Plan only in one sector of forestry *i.e.* reforestation not to mention the employment generated in logging, transportation and processing of other products. It has been found that "with flexibility of many forestry operations, it is often possible to plan labour needs in a way which will complement the agricultural employment pattern and provide much needed jobs in periods of seasonal farm unemployment."

7.15 Types of Employment Provided by Forestry Sector

The forestry sector provided employment which can be classified into following three categories :

(*a*) Direct Employment
(*b*) Self Employment
(*c*) Secondary Employment based on forest industries

However, the secondary employment based on forest industries, like saw-milling, paper, pulp etc. is beyond the scope of this paper because the forestry sector has been defined to include only forest management and logging, collection of minor forest produce etc. Therefore, direct employment and self-employment of the people, generated by forestry sector as defined in Chapter III will only be discussed, without going into details of the employment provided by the industries using forest produce as raw material *e.g.* paper, pulp, saw-milling, manufacture of katha etc.

7.16 Direct Employment in Forestry

The forestry or forest management is a subject which requires knowledge and training in different disciplines. Therefore, direct employment is provided in the form of managerial, technical, research and other executive jobs. Considering the importance of forestry, an All India Service known as Indian Forest Service has been constituted since 1966. The recruitment of the managerial and executive jobs at various levels is done through competitive examination conducted by U.P.S.C. or State Public Service Commission. In addition to the jobs mentioned above the forest administration needs the help of accountants and ministerial staff. The contribution of forestry sector to economic development cannot attain the levels advocated by the planners unless adequate strength of the cadre of professional and technical staff is built up to plan, control, supervise and execute in different forest development and forest management works. The international organizations like F.A.O. are fully aware of the situation and,

Table 7.3

TRAINED MANPOWER REQUIREMENT IN FORESTRY SECTOR OF INDIA IN 1958

Item No.	Activity	Professional level staff	Technicians
1.	Man-made forests	1600	6500
2.	Wood supply	2500	10000
3.	Management of forests	1250	5000
4.	Multiple use forestry	750	3000
5.	Activities—15% of 1, 2, 3 and 4	950	3700
	TOTAL	7050	28200

therefore, they have made appraisal of the situation for underdeveloped countries of Latin America and Africa. Taking inspiration from F.A.O. study, Shri V.K. Seth (1973) has projected the total trained manpower requirements in the forestry sector of the country. The position of trained manpower requirements in the year 1985 based on the works contemplated by the National Commission on Agriculture is as shown in Table 7.3.

As against the requirement of 7000 professionals and 28000 technicians which include Rongers and Foresters, the strength in 1973 was only 1900 and 6000 respectively. This big gap can be bridged only if the training of the professional foresters and technicians is stepped up by setting up more institutes. Efforts in these directions have already been made by different States by setting up more training centres. Even the Government of India has started additional training centres for State Forest Service candidates to cater to the additional requirement of trained forestry personnel. However, a big gap still exists between manpower requirements and availability in the forestry sector. Another category of direct employment provided by the forestry sector consists of the engagement of labour which is recruited locally. These persons are employed in the different forest works under the management and supervision of the professional foresters and technicians. The labourers engaged in different forestry works require almost no skill. All those who reside near or in the forests are employed for forestry works which are of arduous nature and involve more difficult conditions than agricultural or other similar works.

7.17 Self Employment

The self-employment activities based on the forest resource consists of removal of headloads of firewood, fodder from forests, grazing of cattle, collection of minor forest produce like Mahuwa seed, honey, fruits for self-consumption or sale.

7.18 Forest Works Involving Direct Employment

Different activities in forests contributing to the local employment can be broadly grouped as under :

(1) Forest Conservation

(2) Plantation and Nurseries

(3) Fire and Forest Protection

(4) Marking of coupes

(5) Exploitation or harvesting of forest production

(6) Construction and repairs of roads

(7) Construction and repairs of building

(8) Developmental works under Five Year Plans.

7.19 Difficulties in Estimation of Employment in Forest Sector

Most of the forest operations are seasonal and the villagers, men, women and children, go on the work when they have no occupation to supplement their earnings. So far there has been no comprehensive survey in forestry activities. In the forest statistical returns prescribed by the Directorate of Economics and Statistics, Ministry of Agriculture and Irrigation a table is prescribed for reporting employment of labour in forest and forest industries with consumptions of wood and minor forest produce. The table is expected to give the number of persons employed annually (300 mandays are equivalent to the one person year) in (*a*) Management (silviculture) and (*b*) Extraction under broad head forestry. Because of seasonal employment of less than 300 days for a person number of forest labourers is less.

There are numerous forestry operations which are carried out in succession. Some operations are carried out by engaging people on daily wages and other operations are on piece workings. Many a times, the works are got executed through contractors and forest labourers; cooperative societies who directly engage the manual labour on these works. Naturally the record of the employment of workers through agencies other than the Forest Department is not available in the official records. These are the main reasons why it is difficult to assess the employment in forestry and why the total employment in forestry is not adequately reflected in the national statistics or even in the official statistics at district level.

7.20 Methodology of Assessment of the Employment

With full knowledge of the inherent difficulties in assessing the employment in forests, attempts have been made by different forest officers to arrive at the plausible estimate of employment potential and to project the generation of employment in future due to various activities. Seth (1952) estimated the contribution of National Income and working force in the year 1985. The method used by Central Statistical Organisation of Government of India was adopted for Projecting G.N.P. and Gross value of the output in the year 1985. Accordingly, the gross value of the output in forestry has been estimated as Rs. 28000 million. The labour component of this gross value has been taken as 50% *i.e.* Rs. 14000 millions. This value in terms of an average daily wage of Rs. 3, becomes equivalent to nearly 4700 million mandays. And on the basis of seasonal forestry employment of 200 days in a year, this gets converted into the working force figures of 23 million people.

Choudhary (1975) estimated the employment generated in the forestry sector of Chandrapur district for the year 1970-71. In calculating the employment created in mandays, wages paid to the labourers have been adopted for the purpose of the estimation of employment. The employment has been worked out by dividing the total wage component of the expenditure on forest works by the daily wage rate prevailing in the year. According to this method the total number of mandays work out to 8.44 million. In a similar study Parasmis (1976) analysed the role of forestry in rural economy of India with special reference to agriculture in Chandrapur District. Input analysis is carried out by him to study the interrelationship of forestry with agriculture and dairying. In its final analysis, it has been shown that the total number of mandays created by the forestry sector of Chandrapur are about 9.9 million.

The employment potential estimated by Choudhary for the year 1970-71 and employment worked out by Shri Parasnis for the year 1972 have no significant difference. From the discussion made above, it will be clear that the estimation of the employment in terms of mandays is a function of the daily wage rate paid to a labourer and the total expenditure on the labour for a particular operation.

CHAPTER VIII

CONCLUSIONS

8.1 Present Forest Management and Social Change

The importance of the forests as a natural resource and its impact on the environment and its contribution to the income and employment in the rural sector has been described in the earlier chapters. Bulk of these forests in most of the States are managed by the Forest Departments which were established about 100 years ago during the colonical period. In these days, population on the forests was less. The forest management systems were developed in a situation of plenty with the population of India approaching one billion the situation has changed dramatically and the traditional forestry, forest management systems have less relevance. All the same, it is essential that the country's forests are to be managed for productive and environmental functions to achieve human welfare and national development.

8.2 Forest Laws and Rights of the People

No doubt, crores have been invested under various developmental schemes like watershed management, reforestation etc. Unfortunately, these efforts have made little impact on the improvement of the forest cover. With all these efforts it is rather paradoxical that while 23% of the total land area is claimed to be forest area, recent estimates indicate that only about 10% is under good tree cover. The major reason appears to be the lack of public support and involvement mainly because of the lack of attention to the tenure issues and the conflict over the usufruct rights of the rural forest communities. While the rural forest communities are the primary users of the State forests they generally have no formal role to play in the management of the forests. Various forest acts eroded the rights and management system of the communities. The Government has taken the entire responsibility of management of Government forests and also of protection by policing vast areas of forests. The lack of organisational capacity of the forest departments and inability of the local communities to take part in the system because of the laws and regulations regarding forests, have made the management very erratic resulting into overuse of certain forests.

8.3 Constraints in the success of Social Forestry Programmes

Huge investments have been made for the establishment of social forestry programmes in the past. Similar investments are projected in the future also. The main force of the programme was on the plantation of private and community lands which are degraded ecologically. But this programme has not made the desired impact due to lack of legal mechanism to ensure tenure security, right to usufruct etc. The forests departments of the States have tightened their control on the forestry and logging activities on the forest lands by regulating felling and transport of timber. This has been a disincentive to the general social forestry programme on private lands. Thus the lack of proper perspective to deal with fundamental issues regarding rights, forest communities over the forest areas appears to be root cause of all this.

8.4 Need for Change in the Management Pattern

Fortunately, there is a lot of public awareness and concern about forestry, ecology and loss of environment in the recent past. It is upto the Government to harness and use the awareness for the better management and development of the forests not only to arrest the environmental degradation but also to provide sustained supply of goods from the forests to the people dwelling in and around forests.

What is needed is a total change in the attitude and concepts for the forest management and programmes to ensure that direct benefits should go to the people who live in the vicinity of forests. Unless the sense of belongingness to the forest area around the village is developed amongst the village communities, no amount of investment in the reforestation programme or the protection can save the forest from further deterioration causing environmental problems.

The environmental degradation depletes the life support systems and would lead to show erosion in the economic condition of the people, particularly in the rural area.

8.5 Need for the Functional Reclassification of the Government Forests

National Forest Policy of 1952 recommended for classification of forests into (*a*) Protection Forests, (*b*) National Forests, and (*c*) Village Forests. While the protection forests would include national parks, snactuaries, ecologically fragile areas, steep slopes etc., which would need complete protection for the ecological reasons and also for its values regarding bio-diversity, the national forests are productive forests which would provide for the national needs of the forest produce. The village forests of course was for the supply of fodder fuel for the local population on a sustained basis. Unfortunately, this classification could not be achieved. There has been no attempt to assign the forest areas to the individual villages wherefrom the local people would meet their traditional requirement of forest produce.

Considering the future demands, the National Commission on Agriculture (1976) recommended that the forests should be divided into Protection forests, Production forests and Social forests. The Social forests would cover wastelands, Panchayat lands, village commons, road sides, canal banks etc. It has also recommended to include "Minor Forests" which are normally marginal and have been depleted due to over-exploitation and uncontrolled grazing in the past. The Reserved and Protected forests incharge of Forest Department situated on the fringes of villages have depleted and degraded. These forests should be improved by the application of better inputs. No doubt some attempts have been made in this direction. However, there has been little success due to lack of direct association of the village community in the management of these forests. It is, therefore, necessary to carve out certain minimum areas of reserved and protected forests and assign them to the villages without giving them any ownership of land. The villages can be induced to take part in the management and protection of these forests for their economic benefits by showing the produce obtained from these areas.

8.6 People's Participation in Environmental Planning leading to Eco-development

After the forests are assigned to the villages, there is a need to develop a total integrated plan taking village as unit for the eco-development which would cover the total land, water and forest resources of the village. This should aim at the sustained use of these resources for the improvement of the economic conditions of the village communities, with the active participation of the village communities in the management of these resources. Strengthening the community effort is perhaps the only hope for enriching the rural resource base.

8.7 Environment and Sustainable Development

It is very essential to inject environmental dimensions into the whole process of national development and resource planning. The basic objective should be sustainable development. Any ill-planned activity endangers costs which often negate the benefits which are sought for. It has been universally accepted that economic development cannot be achieved at the cost of natural resource of which the forest is the most important renewable resource. The developmental activities like big dams, industries, mining would have a direct impact on the environmental aspects because of the resultant loss of forests. The sustainable development requires co-ordinated planning and implementation to maximise the benefits and to minimise the cost. The urban industrial growth syndrom has driven us to a stage where small easily manageable and worth persuing projects imply do not attract our attention. The case of big dams exposes the very nature of our priorities. Huge excalation of cost during construction, the problem of siltation and reducing the life of the dam, threat to wildlife and environment, rehabilitation of the displaced persons are some of the major problems which are nothing but our own creation. If one looks at their benefits, one is ought to be disappointed. There is little doubt that big dams primarily suit the interests of the industrialists and big agriculturists. The poor which form the majority only bear the brunt of the problem.

India is at the brink of environmental holocaust. Further deterioration would spell doom. And if we keep on denuding out land and be satisfied with ourselves, our reservoirs have limited resources and so is limited the rate of growth. As the rate of consumption overshoots the rate of growth it will result in the depletion of the reservoirs. Therefore, let us change this approach of development which is achieved at the cost of our valuable wealth. It is said that "Old is gold". Old systems of irrigation such as building of wells, tanks and small reservoirs near every farm should be adopted, and also let us concentrate on the slogan, 'a tree for a man'. No doubt we will be able to restore our valuable forest wealth and real economic development.

ANNEXURES

Annexure I

AVAILABILITY OF INDUSTRIAL WOOD AND 'FUELWOOD'

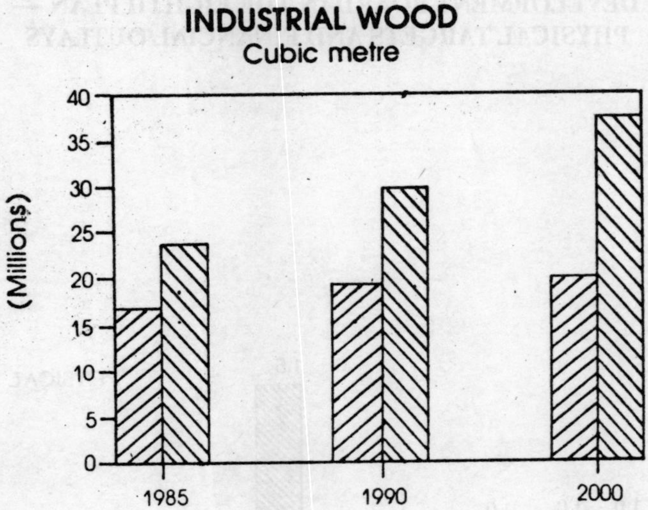

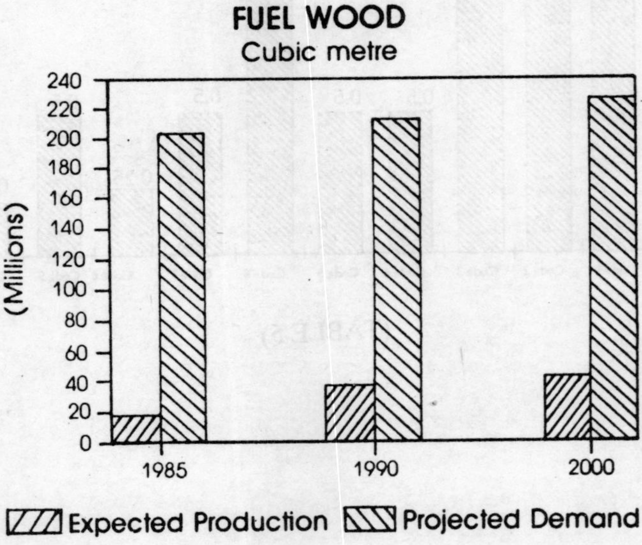

Source : Organisation Plan, Part I.
For Commercial Wood & Fuelwood

Annexure II

SCHEMES TO BE OPERATED BY NATIONAL WASTELANDS DEVELOPMENT BOARD IN THE EIGHTH PLAN — PHYSICAL TARGETS AND FINANCIAL OUTLAYS

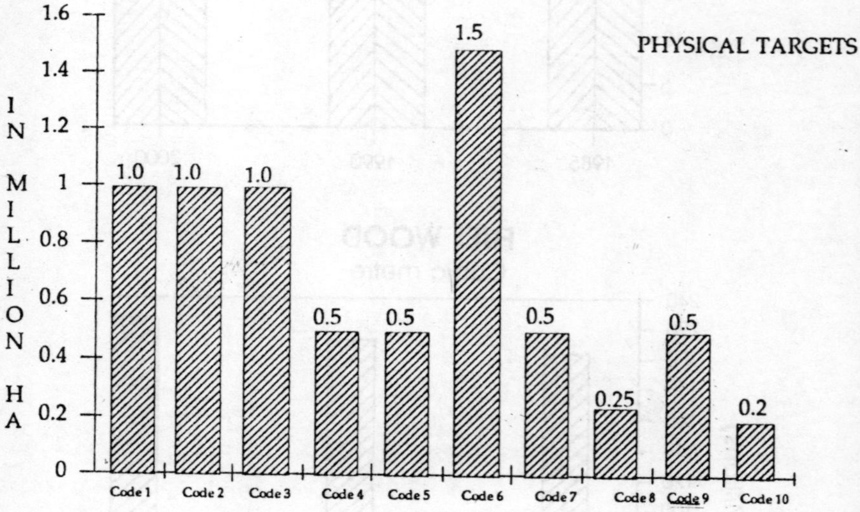

(TABLE 5)

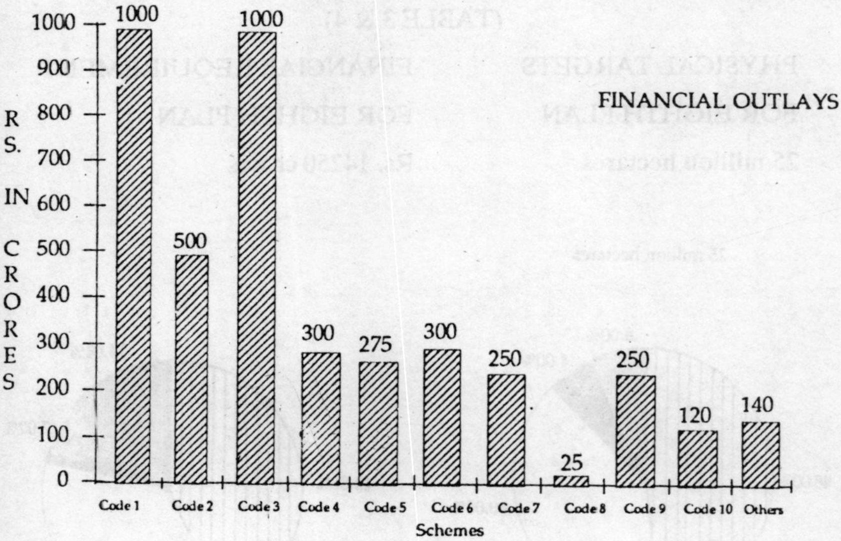

CODE 1	GREENING OF HIMALAYAS.
CODE 2	REGENERATION OF DEGRADED FOREST AREAS.
CODE 3	RECLAMATION OF PROBLEM LANDS.
CODE 4	RURAL FUELWOOD AND FODDER SCHEME.
CODE 5	INTEGRATED WASTELANDS DEVELOPMENT SCHEME.
CODE 6	DECENTRALISED NURSERIES.
CODE 7	MFP — INCLUDING MEDICINAL PLANTS.
CODE 8	AERIAL SEEDING.
CODE 9	MARGIN MONEY ASSISTANCE.
CODE 10	GRANTS-IN-AID TO VOLUNTARY AGENCIES.
	OTHERS :
	SEED DEVELOPMENT AGENCIES.
	PROMOTION OF COMMUNITY STRUCTURE FOR WASTELANDS DEVELOPMENT.
	NWDB HEADQUARTERS.

Annexure III

PHYSICAL TARGETS AND FINANCIAL REQUIREMENTS FOR THE EIGHTH FIVE YEAR PLAN

(TABLE 3 & 4)

PHYSICAL TARGETS FOR EIGHTH PLAN	FINANCIAL REQUIREMENT FOR EIGHTH PLAN
25 million hectares	Rs. 14250 crores

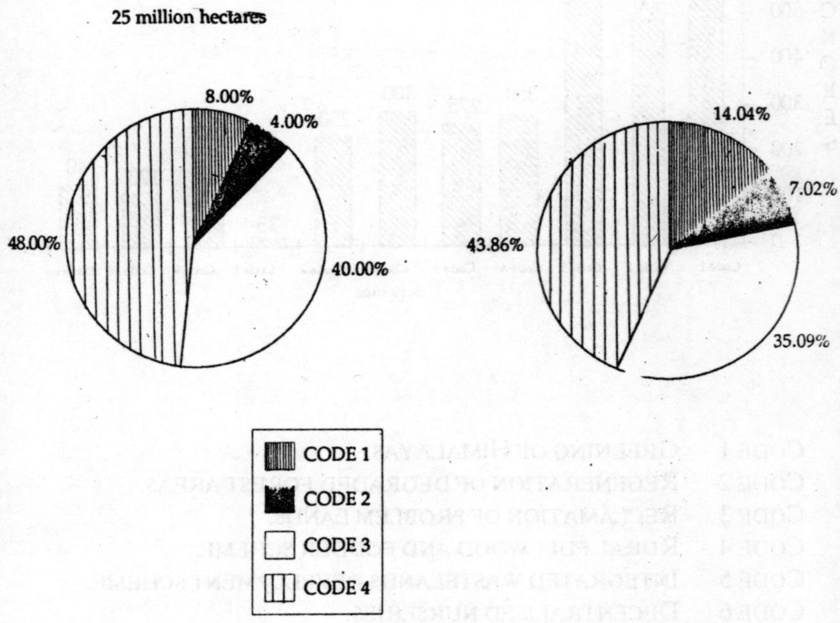

CODE 1 RECLAMATION OF SPECIAL PROBLEM LANDS (TECHNOLOGY EXTENSION)

CODE 2 CONSERVATION OF ECOLOGICALLY FRAGILE HIMALAYAN WATERSHEDS (GREENING OF HIMALAYAS)

CODE 3 REGENERATION OF DEGRADED FOREST AREAS

CODE 4 SOCIAL FORESTRY

Annexure IV

STATEMENT SHOWING THE PHYSICAL AND FINANCIAL ACHIEVEMENTS DURING THE SEVENTH FIVE YEAR PLAN UNDER THE CSS OF ESTABLISHMENT OF SILVIPASTURAL FARMS

S.No.	Name of State	1986-87		1987-88		1988-89	
		Physical (ha.)	Financial (Rs. in lakh)	Physical (ha.)	Financial (Rs. in lakh)	Physical (ha.)	Financial (Rs. in lakh)
(1)	(2)	(3)	(4)	(5)	(6)	(7)	(8)
1.	Andhra Pradesh	—	—	—	—	1000	12.50
2.	Gujarat	—	—	—	—	300 (Advance work)	1.92
3.	Karnataka	83	64.77	74	—	—	—
4.	Orissa	265	4.50	500	3.40	500	48.05
5.	Punjab	—	—	—	—	800	10.00
6.	Rajasthan	—	—	—	—	570 complete 5000 (Advance work)	38.75
7.	Rajasthan (Durgapur)	—	—	—	—	5000	62.50
8.	Uttar Pradesh	—	—	—	—	2080 (Advance work)	26.00
9.	West Bengal	—	—	282	6.75	1560	16.24
		348	69.27	856	10.16		172.715

Annexure V

STATEMENT SHOWING PHYSICAL AND FINANCIAL TARGETS/OUTLAY & ACHIEVEMENTS UNDER THE CSS - DECENTRALISED PEOPLE'S NURSERIES (AS ON 30.9.1988)

S.No.	Name of States	Physical Targers (Seedlings in crore)						Financial Outlay (Rs. in lakhs)					
		1986-87		1987-88		1988-89		1986-87		1987-88		1988-89	
		Tar.	Ach.	Tar.	Ach.	Tar.	Ach.	Allo-cation	Funds release	Allo-cation	Funds release	Allo-cation	Funds release
1	2	3	4	5	6	7	8	9	10	11	12	13	14
1.	Andhra Pradesh	0.22	NIL	1.33	NIL	1.33	—	10.00	10.00	60.00	—	50.00	22.50
2.	Arunachal Pradesh	—	—	—	—	0.04	—	—	—	—	—	2.00	1.00
3.	Assam	0.22	0.22	0.44	0.23	0.44	—	10.00	10.00	20.00	10.00	20.00	9.00
4.	Gujarat	4.44	9.20	2.22 + 5 crore Salva-ging	2.70 + 5 crore Salva-ging	3.60	—	200.00	200.00	200.00	198.75	160.00	187.00
5.	Haryana	1.33	1.33	1.33	1.30	1.14	—	60.00	60.00	60.00	—	50.00	—
6.	Himachal Pradesh	0.44	NIL	0.55	0.62	—	—	20.00	20.00	25.00	—	—	—
7.	Jammu & Kashmir	0.44	0.10	0.34	NR	—	—	20.00	20.00	15.10	—	—	—
8.	Kerala	1.11	NIL	2.22	NIL	—	—	50.00	50.00	100.00	—	—	—
9.	Karnataka	8.48	7.87*	10.00	7.14	3.78	—	—	—	450.00	150.00	170.00	245.86
10.	Madhya Pradesh	2.21	1.80	6.50	5.55	2.28	—	100.00	100.00	292.50	111.00	102.50	420.50
11.	Maharashtra	3.32	4.55*	7.00	5.55	3.33	—	127.00	127.00	315.00	110.00	150.00	137.00

12.	Manipur	0.22	—	0.22	—	0.22	—	10.00	10.00	10.00	—	—	—
13.	Meghalaya	0.22	0.16	0.22	0.22	0.22	—	10.00	10.00	10.00	10.00	10.00	9.50
14.	Mizoram	0.22	0.26	0.44	0.28	0.44	—	10.00	10.00	20.00	7.20	20.00	9.00
15.	Nagaland	0.22	NIL	0.44	0.19	0.22	—	10.00	10.00	20.00	—	10.00	10.00
16.	Orissa	1.33	1.50	1.33	0.67	1.11	—	60.00	60.00	60.00	27.23	50.00	50.00
17.	Punjab	0.88	NIL	1.33	0.89	1.32	—	40.00	40.00	60.00	20.00	40.00	18.00
18.	Rajasthan	1.67	1.00*	1.11	0.33	2.11	—	30.00	20.00	50.00	—	80.00	35.50
19.	Sikkim	—	—	—	—	0.04	—	—	—	—	—	2.00	1.00
20.	Tamil Nadu	0.44	NR	0.22	0.62	—	—	20.00	20.00	10.00	—	—	97.50
21.	Tripura	0.22	0.06	0.22	0.61	0.22	—	10.00	10.00	10.00	5.00	10.00	4.50
22.	West Bengal	1.55	2.33	2.22	1.55	1.11	—	70.00	70.00	100.00	74.65	50.00	101.74
23.	Delhi	0.06	NR	0.11	NR	—	—	3.00	3.00	5.00	—	—	—
24.	Bihar	—	—	—	—	1.00	—	—	—	—	—	—	45.00
	Total	29.24	30.38	39.80 + 5 crore Salvaging	28.45	22.95	—	870.00	870.00	1892.60	723.83	976.50 +323.50	1461.01
	Others NDDB	2.76	1.78					124.29	124.29			1300.00	
	Grand Total	32.00	32.16		28.45			994.29	994.29				

* Indicates progress made in respect of the funds released as on 31.3.1986 as indicated below:

1. Karnataka 346.50 lakhs
2. Rajasthan 45.00 lakhs
3. Maharashtra 22.50 lakhs

Total 414.00 lakhs

Annexure VI

NATIONAL WASTELAND DEVELOPMENT BOARD : GRANTS-IN-AID SCHEME TO VOLUNTARY AGENCIES

State-wise amount sanctioned, released and Number of projects as on 23/02/89

State	1985-86			1986-87			1987-88		
	Amount Sanctioned	Amount Released	No. of Projects	Amount Sanctioned	Amount Released	No. of Projects	Amount Sanctioned	Amount Released	No. of Projects
Andhra Pradesh	6173641.00	6173641.00	4	62531701.00	3652059.50	14	4151870.00	17217651.00	8
Assam	0.00	0.00	0	0.00	0.00	0	230000.00	230000.00	1
Bihar	406500.00	276750.00	1	1948546.50	1603421.50	7	5935027.00	2622666.00	10
Goa	0.00	0.00	0	0.00	0.00	0	5995.00	4196.50	1
Gujarat	7893375.00	2725926.00	2	12307095.15	6539104.55	9	11515517.00	7359903.00	13
Haryana	0.00	0.00	0	1545150.00	373250.00	4	0.00	0.00	0
Himachal Pradesh	0.00	0.00	0	0.00	0.00	0	73550.00	30000.00	1
Karnataka	0.00	0.00	0	2225361.20	1807497.20	4	315239.00	188689.00	4
Kerala	0.00	0.00	0	0.00	0.00	0	584860.00	340660.00	2
Madhya Pradesh	107000.00	107000.00	1	3272066.00	2058948.00	5	21500000.00	3000000.00	1
Maharashtra	12000.00	12000.00	1	6514463.75	2984564.76	14	6079070.00	4460030.00	12
Manipur	101210.00	101210.00	1	184940.00	120553.35	1	1480815.00	405513.00	4
Orissa	588500.00	588500.00	2	3015336.50	793836.50	4	1474175.00	269375.00	4
Rajasthan	3716740.00	3716740.00	4	16737846.25	9665586.25	8	29382957.00	8563217.00	18
Tamil Nadu	48200.00	48200.00	1	10067965.00	3275365.00	9	4739591.25	4259992.25	14
Uttar Pradesh	1523204.00	536804.00	4	4333500.00	2846195.00	8	723875.00	340213.00	3
West Bengal	0.00	0.00	0	3552119.00	1312740.00	8	2929180.00	1230431.00	4
Delhi	3510450.00	3510450.00	2	875945.00	340315.00	2	107000.00	74900.00	2
Grand Total	18524543.00	12240944.00	23	72833504.35	37373436.61	97	91228721.25	35101550.75	102

(Contd.)

State	1988-89 Amount Sanctioned	Amount Released	No. of Projects	Total: Total Amount Sanctioned	Total Amount Released	Total Projects
Andhra Pradesh	270125.00	3246168.00	2	11292529.00	9237356.50	28
Assam	0.00	0.00	0	23000.00	23000.00	1
Bihar	4830125.00	4082518.00	6	13120198.50	8585355.50	24
Goa	0.00	0.00	0	5995.00	4196.50	1
Gujarat	12881450.00	9786657.00	7	44597437.15	26411590.55	31
Haryana	30000.00	30000.00	1	1575150.00	403250.00	5
Himachal Pradesh	0.00	43550.00	0	73550.00	73550.00	1
Karnataka	1383600.00	1501676.00	2	3924200.20	3497862.20	10
Kerala	1509600.00	654747.00	2	2094460.00	995407.00	4
Madhya Pradesh	0.00	0.00	0	24879066.00	5165948.00	7
Maharashtra	5041624.00	4213195.00	4	17647157.75	11669789.76	31
Manipur	502840.00	430565.00	2	2269805.00	1057841.35	8
Orissa	325530.00	120800.00	1	5403541.50	1772511.50	11
Rajasthan	2943750.00	5299299.00	1	52781293.25	27244842.25	31
Tamil Nadu	2455841.00	3187914.44	9	17311597.25	10771471.69	33
Uttar Pradesh	316250.00	860640.49	1	6896829.00	4583852.49	16
West Bengal	19307070.00	976199.00	11	8412006.00	3519370.00	23
Delhi	0.00	0.00	0	4493395.00	3925665.00	6
Grand Total	34421442.00	34433928.93	49	217008210.60	119149860.29	271

Annexure VII

PHYSICAL AND FINANCIAL COMPONENTS OF EXTERNALLY AIDED SOCIAL FORESTRY PROJECTS

Financial (F) : Rs. in Millions
Physical (P) : In Hectares

S.No.	Name of the Project	Donor Agency	Project Period	Project Cost		Project Components						
						Farm Forestry Project	Village Woodlots	Strip Plantations	Ref/ Ref of Degraded Forests	Support Activities	Organisational Functions	TOTAL
1	2	3	4	5		6	7	8	9	10	11	12
1	National Social Forestry Project Uttar Pradesh	World Bank USAID	5 Years 1985-86 to 1989-90	1,611.60	P	147,210	14,000	740	—	—	—	161,950
					F	536.20	150.40	17.20	—	216.60	641.20	1,611.60
2	National Social Forestry Project Gujarat	World Bank USAID	5 Years 1985-86 to 1989-90	1,296.50	P	230,500	35,000	17,500	30,400	—	—	313,400
					F	143.60	369.80	630.80		32.40	119.90	1,296.50
3	National Social Forestry Project Himachal Pradesh	World Bank USAID	5 Years 1985-86 to 1989-90	572.90	P	66,838	41,000	—	5,000	—	—	112,833
					F	139.30	201.20	—	24.60	49.30	158.50	572.90
4	National Social Forestry Project Rajasthan	World Bank USAID	5 Years 1985-86 to 1989-90	391.90	P	91,500	5,000	4,300	20,000	—	—	120,800
					F	121.80	30.20	113.10		37.30	89.50	391.90

5	Haryana Social Forestry Project	World Bank DANIDA	1982-83 to 1986-87 extended to 1989-90	333.25	P	30,000	12,000	9,500	15,500	—	—	67,000
					F	25.65	46.39	45.35	43.39	9.80	88.30 +74.37 Price escalation & Contingencies	333.25
6	Jammu and Kashmir Social Forestry Project	World Bank DANIDA	1982-83 to 1986-87 extended to	237.20	P	19,000	5,000	3,000	17,000	—	—	44,000
					F	17.40	17.00	11.70	64.20	14.50	+53.80 & Contingencies	237.40
7	Karnataka Social Forestry Project	World Bank ODA (U.K.)	1983-84 to 1987-88	552.30	P	120,500	25,000	4,000	—	—	—	149,500
					F	180.00	98.50	9.30	—	8.30	98.50 Price escalation & Contingencies	552.30
8	Kerala Social Forestry Project	World Bank	1984-85 to 1989-90	599.11	P	69,200	14,100	2,000	—	—	—	85,300
					F	162.25	183.43		—	18.77	234.66	599.11
9	West Bengal Social Forestry Project	World Bank	1981-82 to 1986-87 extended to 1989-90	348.65	P	52,000	6,000	20,000	15,000	—	—	93,000
					F	18.84	17.84	91.47	51.04	7.38	72.84 +89.24 Price escalation & Contingencies	348.65
10	Bihar Social Forestry Project	SIDA	1985-86 to 1990-91	538.57	P	71,750	30,750	1,200	64,500	—	—	168,200
					F	156.28	21.52	18.00	131.30	16.00	195.47	538.57

1	2	3	4	5	6	7	8	9	10	11	12
11	Orissa Social Forestry Project										
	Phase 1	SIDA	1983-84 to	281.70	P 26,500	21,700		35,300	—	—	134,400
			1987-88		F 11.60		170.70		11.65	87.75	281.70
	Phase II	SIDA	1988-89 to	783.40	P 62,000	52,500	650	19,250	—	—	83,500
			1992-93		F 253.60	222.30	220	65.60	41.50	198.20	783.40
12	Tamil Nadu Social Forestry										
	Project Phase I	SIDA	1981-82 to	591.38	P 85,165	131,405	7,925	—	—	—	224,495
			1985-86 extended to 1987-88	F	50.80 178.94	39.46	170.94	18.63	132.61	591.38	
	Phase II	SIDA	1988-89 to	854.00	P 18,000	65,300	—	4,080	—	—	78,380
			1992-93		F 154.40	247.50	—	25.80	45.50	260.40 +120.40 (Unallo-cated)	854.00
13	Andhra Pradesh	CIDA	1983-84 to		P 108,100	25,000	3,785	13,840	—	—	150,725
	Social Forestry		1987-88	383.78							
	Project		extended to 1989-90		F 68.78	61.03	62.28	82.45	11.86	97.38	388.78
14	Maharashtra	USAID	1982-83 to								
	Social Forestry		1989-90	564.00	P 44,035	33,975	2,990	—	—	—	81,000
	Project				F 257.74			—	47.01	259.25	564.00
	TOTAL				P1,242,293	508,730	77,590	239,870	—	—	2,068,483
				F 9940.44	F 2,090.50		3,971.67		586.50	3,288.77	F 9940.44

Note:
1. Farm Forestry Component includes National Conversion @ 2000 seedlings/ha for Seedlings distribution.
2. The totals relate to period 1981-82 to 1992-93.

Annexure VIII

IMPORTANT FORESTRY SCHEMES IN THE STATE PLAN SECTOR

1. Forestry research, education and training
2. Forest Conservation and Development
   - (*i*) Forest protection
   - (*ii*) Fire protection
   - (*iii*) Forest protection force
   - (*iv*) Development forest villages
   - (*v*) Development of infrastructure for the protection of forests from biotic interference
3. Forest Resources Survey
   - (*i*) Working plan preparations
   - (*ii*) Consolidation and demarcation
   - (*iii*) Survey of minor forest produce
   - (*iv*) Forest settlement operations
4. Extension and publicity
5. World food programme
6. Indo-German Dhaulandhar Project
7. Rehabilitation of degraded forests
8. Investment in public sector and other undertakings
9. Project formulation, monitoring and evaluation
10. District councils secretariat.
11. Soil conservation
12. Development of minor forest produce
13. Communication (roads) and Buildings
14. Afforestation schemes
   - (*i*) Rural fuelwood plantations
   - (*ii*) Mechanised plantations

(*iii*) Economic plantations

(*iv*) Social forestry projects (externally aided)

(*v*) Silvipastoral farms

(*vi*) Road sites plantations

(*vii*) Social forestry in urban areas

(*viii*) Energy plantations

(*ix*) Quick growing species

(*x*) Area oriented schemes for fuelwood and fodder

15. Integrated ravine reclamation in the catchment of the river Chambal

16. Environmental forestry and wildlife

(*i*) Intensive management of wildlife sancturies and national parks

(*ii*) Tiger reserves

(*iii*) Assistance for captive breeding and rehabilitation of endangered species

(*iv*) Development of zoos

(*v*) Nature education and interpretation programme

(*vi*) Assistance for the control of poaching and illegal trade.

Annexure IX

ALLOCATION OF FUNDS AND AREA COVERED FOR AFFORESTATION UNDER RDD SCHEMES

	Year	Funds Allocated (Rs. in lakhs)	Funds Utilised (Rs. in lakhs)	Area Covered (ha)
NREP	1985-86	9149.20	7740.05	115553
	1986-87	11435.00	11638.99	210589
	1987-88	13199.50	10893.50	139986
	1988-89	16455.50	N.A.	N.A.
		50239.20	30272.54	466128
RLEGP	1985-86	6380.40	4801.35	53300
	1986-87	9250.00	9624.16	219144
	1987-88	8825.00	5097.49	162160
	1988-89	8133.00	N.A.	N.A.
		32588.40	19523.00	434604
DPAP	1985-86	2129.62	1642.24	52298
	1986-87	2650.54	2034.93	57845
	1987-88	2306.25	2007.32	72218
	1988-89	2555.76	N.A.	34574 (up to 9/88)
		9642.17	5684.49	216935
DDP	1985-86	687.31	733.26	18654
	1986-87	1288.66	1391.89	13231
	1987-88	1296.92	1210.23	17260
	1988-89	1980.00	N.A.	11622 (up to 9/88)
		5252.89	3335.38	60767
Grand Total		97722.00	58815.41	1178434

Annexure X

STATEMENT SHOWING PROGRESS OF AFFORESTATION THROUGH SUCCESSIVE PLANS

		Physical		Financial	
Sl.No.	Five Year Plan Period	Area Afforested in Plan Period (Lakh ha.)	Cumulative (Lakh ha.)	Afforestation Expenditure in Plan (Rs. in crores)	Cumulative Expenditure (Rs. in crores)
(1)	(2)	(3)	(4)	(5)	(6)
1.	First	0.52	0.52	1.28	1.28
2.	Second	3.11	3.63	6.86	8.14
3.	Third	5.83	9.46	21.13	29.27
4.	1966-69	4.53	13.99	23.02	52.29
5.	Fourth	7.14	21.13	44.34	96.63
6.	Fifth	12.21	33.34	107.28	203.91
7.	1979-80	2.22	25.56	37.10	241.01
8.	Sixth	46.50	82.06	926.01	1167.02
9.	Seventh				
	(*a*) 1985-86	15.10	97.16	374.58	1541.60
	(*b*) 1986-87	17.62	114.78	491.74	2033.34
	(*c*) 1987-88	17.75	132.53	*	*

* Final Figures Awaited

Annexure XI

SHARE OF FARM FORESTRY IN TOTAL AFFORESTATION (SEEDLING IN CRORES)

Year	Farm Forestry	Total	% (2 To 3)
1981-82	44.1	131.8	33%
1982-83	88.7	207.24	41%
1983-84	103.7	241.2	41%
1984-85	128.0	264.0	49%
1985-86	139.0	302.0	46%
1986-87	160.0	352.0	46%
1987-88	175.0	355.0	49%
1988-89	200.0	400.0	50%

Annexure XII

STATEWISE PLAN EXPENDITURE ON AFFORESTATION FROM 1985-86 TO 1987-88

(Rs. in lakhs)

Sl.No.	States/Uts	Expenditure		Allocations*	
		1985-86	1986-87	1987-88	1988-89
1.	Andhra Pradesh	2538.01	3290.56	3022.70	3538.00
2.	Assam	1664.77	1799.93	1784.50	2188.00
3.	Bihar	3129.01	4048.92	3281.00	5298.00
4.	Gujarat	2280.81	2936.94	2859.25	3168.00
5.	Haryana	1316.62	1327.34	1460.80	1921.50
6.	Himachal Pradesh	1232.99	1637.53	1919.50	2257.50
7.	Jammu & Kashmir	659.39	843.56	970.25	1124.63
8.	Karnataka	2276.37	2076.83	2085.00	2710.50
9.	Kerala	1528.40	1717.50	1812.75	2374.00
10.	Madhya Pradesh	3234.95	4279.89	4238.25	4672.00
11.	Maharashtra	2896.82	4032.69	3663.55	4194.25
12.	Manipur	224.25	249.40	306.50	703.50
13.	Meghalaya	522.48	521.88	670.75	756.00
14.	Nagaland	351.14	401.96	519.00	518.00
15.	Orissa	1536.06	1985.50	2416.45	2667.25
16.	Punjab	747.56	853.40	851.00	1035.25
17.	Rajasthan	1545.99	2892.17	2741.75	3202.00
18.	Sikkim	182.98	188.78	228.25	235.00
19.	Tamil Nadu	2738.96	3243.10	3141.25	3479.50
20.	Tripura	389.91	416.90	434.50	462.75
21.	Uttar Pradesh	3659.33	6443.28	6191.00	7589.75
22.	West Bengal	2135.46	2574.08	2228.20	3292.88
23.	A & N Islands	136.00	122.0	143.50	259.50
24.	Arunachal Pradesh	374.60	399.00	438.00	451.75
25.	Chandigarh	28.60	23.25	24.35	23.50
26.	D.N. Haveli	56.80	46.50	68.25	111.25
27.	Delhi	67.90	80.75	113.80	85.50
28.	Goa Daman Diu	106.70	114.50	139.65	163.00
29.	Mizoram	368.36	537.59	4.75	658.00
30.	Pondicherry	22.60	31.00	527.50	48.00
31.	Lakshadweep	3.60	4.25	38.50	7.25
	Total	37858.02	49120.98	48326.55	59196.01

* Anticipated

Annexure XIII

STATE-WISE AREA COVERAGE UNDER AFFORESTATION PROGRAMME—1985-86 TO 1988-89

(Area in hectares)

Sl.No.	State/UTs	1985-86	1986-87	1987-88	1988-89 (Target)
1.	Andhra Pradesh	157800.00	143707.50	15267.00	160000
2.	Assam	19800.00	31275.00	24893.50	30000
3.	Bihar	76150.	135550.00	157600.00	180000
4.	Gujarat	124850.00	113550.	107075.00	130000
5.	Haryana	46850.00	37079.00	19000.00	37500
6.	Himachal Pradesh	33600.00	33564.00	30754.50	35000
7.	Jammu & Kashmir	23350.00	28526.00	20000.03	25000
8.	Karnataka	127300.00	115837.00	157610.00	165000
9.	Kerala	58300.00	75962.00	77772.00	87500
10.	Madhya Pradesh	175050.00	196000.00	204523.00	220000
11.	Maharashtra	108250.00	119085.00	153998.00	165000
12.	Manipur	6250.00	7440.00	9012.50	10000
13.	Meghalaya	6550.00	7900.00	11878.50	13500
14.	Nagaland	13450.00	27175.00	10000.00	11500
15.	Orissa	96500	116336.00	117002.00	150000
16.	Punjab	29500.00	28379.50	24776.00	25000
17.	Rajasthan	47900.00	67051.50	58693.50	65000
18.	Sikkim	4100.00	5751.50	6693.50	7500
19.	Tamil Nadu	60750.00	99064.00	95587.00	90000
20.	Tripura	10000.00	13150.00	13356.50	13000
21.	Uttar Pradesh	177400.00	243250.00	221035.50	255000
22.	West Bengal	55750.00	70800	69554.00	90000
23.	A & N Islands	4750.00	6116.00	5021.50	5000
24.	Arunachal Pradesh	5150.00	6254.00	6352.00	7000
25.	Chandigarh	76.00	191.50	179.50	200
26.	D & N. Haveli	1550.00	1700.00	1561.00	1750
27.	Delhi	1250.00	3151.50	903.00	2500
28.	Goa Daman Diu	2250.00	3396.50	3761.50	3050
29.	Lakshadweep	12.50	14.50	12.	25
30.	Mizoram	35000.00	23902.50	13875.00	15000
31.	Pondicherry	550.00	649.50	516.00	300
All India Total:		1510038.50	1761869.00	1775564.03	2001325

Area notionally computed @ 2000 saplings per hectare.

Annexure XIV

PROGRESS OF AFFORESTATION UNDER TWENTY POINT PROGRAMME—POINT NO. 16 : TREE PLANTATION (1989-89)

State-wise Targets & Achievements (1988-89)

(No. of Seedlings in Lakhs)

S.No.	States/U.Ts.	Target (1988-89)	Achievement Up to Dec.88.
1.	Andhra Pradesh	3200	2570.87
2.	Arunachal Pradesh	140	31.44
3.	Assam	600	445.50
4.	Bihar	3600	3223.00
5.	Goa	75	71.98
6.	Gujarat	2600	4013.61
7.	Haryana	750	538.38
8.	Himachal Pradesh	700	590.28
9.	Jammu & Kashmir	500	159.03
10.	Karnataka	3300	2872.02
11.	Kerala	1750	1352.55
12.	Madhya Pradesh	4400	4241.00
13.	Maharashtra	3300	4028.79
14.	Manipur	200	198.96
15.	Meghalaya	270	329.77
16.	Mizoram	300	300.00
17.	Nagaland	230	230.00
18.	Orissa	3000	2750.12
19.	Punjab	500	472.19
20.	Rajasthan	1300	1259.00
21.	Sikkim	150	111.35
22.	Tamil Nadu	1800	1320.73
23.	Tripura	260	257.00
24.	Uttar Pradesh	5100	5311.95
25.	West Bengal	1800	1100.00
26.	A & N Islands	100	105.63
27.	Chandigarh	4	3.21
28.	D & N Haveli	35	38.32
29.	Daman & Diu	2	8.90
30.	Delhi	50	57.62
31.	Lakshadweep	0.50	2.21
32.	Pondicherry	10	5.62
	Grand Total	40026.50	38003.03
	@ 2000 Trees/Ha.	2.0 M.H.	1.90 M.Ha.

Annexure XVA

STATEMENT SHOWING STATE-WISE AND YEAR-WISE TOTAL PHYSICAL FINANCIAL ACHIEVEMENT/EXPENDITURES UNDER RURAL FUELWOOD PLANTATION SCHEME FROM 1980-81 TO 1983-84

Physical : (1) Plantation—in ha.
(2) Distribution of seedlings—in lakh nos.

Financial : Rs. in lakhs

S.No.	State/UT	1980-81			1981-82			1982-83			1983-84		
		Plan-. tation raised	Seedling distri- buted	Total expen- diture	Plan- tation raised	Seedling distri- buted	Total expen- diture	Plan- tation raised	Seedling distri- buted	Total expen- diture	Plan- tation raised	Seedling distri- buted	Total Expen- diture
1	2	3	4	5	6	7	8	9	10	11	12	13	14
1.	Andhra Pradesh	—	—	9.51	4446	133.88	121.88	5413	166.07	150.04	5528	179.70	175.85
2.	Assam	—	—	—	1450	40.05	50.18	2824	62.58	78.56	4600	66.95	135.58
3.	Bihar	—	—	2.14	335	—	72.65	7000	106.00	175.91	7000	71.76	200.00
4.	Gujarat	—	—	21.43	2000	150.00	71.41	2400	164.00	113.45	3000	197.00	152.64
5.	Haryana	—	—	—	3326	40.00	113.89	2500	62.50	98.95	2500	62.50	86.15
6.	Himachal Pradesh	241	—	8.01	1349	4.00	32.39	2054	40.00	72.43	3480	63.76	132.42
7.	Jammu & Kashmir	—	—	—	—	—	—	—	—	—	—	—	—

1	2	3	4	5	6	7	8	9	10	11	12	13	14
8.	Karnataka	104	102.67	23.73	3144	75.00	111.47	3328	200.00	132.58	4462	175.00	173.08
9.	Kerala	1222	9.00	43.12	1365	34.47	74.55	5327	80.51	86.09	7561	80.03	72.62
10.	Madhya Pradesh	—	—	97.38	3273	210.68	268.62	10695	286.49	395.92	12135	297.50	286.34
11.	Maharashtra	—	—	—	—	18.30	17.68	1019	25.80	58.94	1726	171.42	88.73
12.	Manipur	—	—	—	—	4.00	7.95	520	5.00	10.56	600	6.00	19.72
13.	Maghalaya	—	—	1.39	439	12.00	13.91	821	12.00	21.15	500	1.27	14.76
14.	Nagaland	—	—	9.48	948	3.80	24.03	1481	23.45	36.64	2400	33.00	72.37
15.	Orissa	—	—	8.42	3839	6.10	49.85	6411	61.00	108.62	7308	73.94	118.96
16.	Punjab	—	—	—	2500	60.24	69.84	2500	90.10	79.88	3006	100.14	134.00
17.	Rajasthan	—	—	50.84	4008	—	89.84	3900	20.00	165.36	7800	100.50	202.23
18.	Sikkim	96	—	1.92	230	5.00	7.09	334	5.00	9.17	305	5.00	9.90
19.	Tamil Nadu	—	—	11.61	7861	10.00	81.27	6500	37.60	104.13	5179	111.37	130.84
20.	Tripura	—	—	—	227	6.00	4.22	1080	7.50	10.28	1523	20.00	20.00
21.	Uttar Pradesh	—	—	—	1525	—	69.41	3408	—	72.16	3643	—	147.66
22.	West Bengal	—	—	6.42	685	—	30.68	1750	50.00	48.71	1780	58.00	51.00
23.	Arunachal Pradesh	—	—	1.31	90	0.50	5.98	510	0.87	9.07	1036	0.65	22.09
24.	Mizoram	1500	0.40	12.10	600	2.00	36.10	4000	4.00	48.80	4000	4.00	71.91
25.	Goa	—	—	—	—	—	—	—	—	—	140	1.50	2.50
26.	Delhi	—	—	—	—	—	—	—	—	—	150	1.81	0.47
	Total	3163	112.07	308.81	43640	816.02	1424.89	75775	1510.47	2087.40	91362	1882.80	2521.92

Annexure XVB

STATEMENT SHOWING PHYSICAL/FINANCIAL ACHIEVEMENTS MADE FROM 1980-81 TO 1988-89 UNDER THE CENTRALLY SPONSORED SCHEME OF SOIL, WATER AND TREE CONSERVATION IN THE HIMALAYAS (OPERATION SOILWATCH)

Physical : In Ha. ; Financial : Rs. in lakhs

(1) 1980-81 to 1984-85

S.No.	Name of State	1980-81		1981-82		1982-83		1983-84		1984-85	
		Area treated	Expenditure incurred	Area treated	Expenditure incurred	Area treated	Expenditure incurred	Area treated	Expenditure incurred	Area treated	Expenditure incurred
1.	Assam	—	—	—	—	—	—	—	—	950	32.33
2.	Haryana	790	15.78	1955	36.18	1440	74.11	1772	79.76	2495	140.00
3.	Himachal Pradesh	1701	91.85	3580	110.06	3807	124.44	6182	147.92	7874	201.99
4.	Jammu & Kashmir	—	2.42	189	5.58	330	10.45	510	19.20	1083	34.56
5.	Manipur	—	—	—	—	—	—	95	19.93	1080	30.00
6.	Meghalaya	—	—	—	—	—	—	—	—	2160	40.01

(Contd.)

S.No.	Name of State	1980-81		1981-82		1982-83		1983-84		1984-85	
		Area treated	Expenditure incurred	Area treated	Expenditure incurred	Area treated	Expenditure incurred	Area treated	Expenditure incurred	Area treated	Expenditure incurred
7.	Nagaland	—	13.37	1171	20.94	1327	29.95	1550	45.89	2400	65.84
8.	Punjab	—	20.00	3500	50.00	2560	50.00	2500	50.00	2500	62.16
9.	Sikkim	617	15.15	295	9.74	310	12.00	600	20.00	850	30.68
10.	Tripura	—	—	1085	12.99	1139	20.12	1296	18.94	1365	24.32
11.	Uttar Pradesh	6495	145.65	7593	200.76	5703	161.91	7663	159.58	7562	219.43
12.	West Bengal	82	13.96	128	13.56	153	17.34	190	38.61	688	66.37
13.	Arunachal Pradesh	270	13.27	262	13.85	247	11.64	270	11.65	264	10.92
14.	Mizoram	1325	15.02	1720	21.44	1750	30.30	1195	52.11	3250	70.00
	Total	11280	346.47	21478	495.10	18766	524.26	24423	663.59	34521	1028.70

(2) 1985-86 to 1988-89

S.No.	Name of State	1985-86		1986-87		1987-88		1988-89	
		Area treated	Expenditure incurred	Area treated	Expenditure incurred	Area treated	Expenditure incurred	Target	Financial outlay
1.	Assam	336	9.51	683	27.61	1000	39.92	1000	42.50
2.	Haryana	3912	195.67	4450	158.50	3593	150.15	2460	120.22
3.	Himachal Pradesh	4146	225.45	5505	220.46	5274	211.24	4650	220.20
4.	Jammu & Kashmir	861	32.44	1003	35.57	1956	74.94	1785	80.00
5.	Manipur	1385	41.50	1564	39.20	1080	43.40	1980	43.82
6.	Meghalaya	1751	59.48	3358	74.18	3060	93.58	4011	82.30
7.	Nagaland	3600	101.82	4500	110.23	6000	121.25	2350	90.00
8.	Punjab	3750	75.00	3750	121.66	2280	87.78	2270	100.00
9.	Sikkim	919	38.16	1240	45.00	1042	43.32	903	40.00
10.	Tripura	1271	27.60	1322	31.50	1253	30.84	1000	40.00

(Contd.)

S.No.	Name of State	1985-86		1986-87		1987-88		1988-89	
		Area treated	Expenditure incurred	Area treated	Expenditure incurred	Area treated	Expenditure incurred	Target	Financial outlay
11.	Uttar Pradesh	5709	192.48	4291	162.54	2532	119.41	3427	160.00
12.	West Bengal	900	69.91	1764	69.41	1804	79.18	1539	73.56
13.	Arunachal Pradesh	239	4.46	114	1.80	70	1.15	50	1.20
14.	Mizoram	3200	82.50	4011	112.00	1790	113.00	1750	63.48
	Total	31979	1155.98	37555	1212.66	32734	1209.16	29175	1157.28

Annexure XVC

STATEMENT SHOWING STATE-WISE AND YEAR-WISE TOTAL PHYSICAL FINANCIAL ACHIEVEMENT/EXPENDITURES UNDER RURAL FUEL WOOD PLANTATION SCHEME FROM 1984-85 TO 1988-89

Physical : (1) Plantation — in Ha.
(2) Distribution of seedlings — in lakh nos.

Financial : Rs. in lakhs

S.No.	State/UT	1984-85			1985-86			1986-87			1987-88		1988-89	
		Plantation raised	Seedling Distributed	Total expenditure	Plantation raised	Seedling distributed	Total expenditure	Plantation raised	Seedling distributed	Total expenditure	Plantation raised	Total expenditure	Plantation raised	Outlay total
1.	Andhra Pradesh	4823	237.07	216.70	3329	384.50	177.08	3590	—	216.90	2250	184.73	4100	200.00
2.	Assam	4600	160.00	182.35	5200	174.00	260.26	5200	97.40	234.54	3400	185.89	5153	200.00
3.	Bihar	7870	8.60	230.00	8300	196.15	300.00	9600	150.00	290.00	9225	350.00	10516	300
4.	Gujarat	3500	245.00	212.39	3588	245.00	184.18	2564	25.00	163.17	1446	137.38	4839	232.41
5.	Haryana	2500	62.50	103.64	2000	84.19	109.75	2450	—	127.45	1700	134.02	2000	135.00
6.	Himachal Pradesh	3810	36.58	192.30	4226	—	150.63	3851	—	178.20	2594	167.19	2070	185.03
7.	Jammu & Kashmir	—	—	—	1150	—	28.39	1115	10.40	33.08	1314	70.85	1380	80.00
8.	Karnataka	3700	200.00	250.56	266	120.00	192.55	1137	120.00	160.40	3646	178.51	2529	241.49
9.	Kerala	3985	100.00	93.71	3627	75.85	116.45	7	55.56	16.09	1628	53.94	1500	120.00
10.	Madhya Pradesh	8892	348.25	304.26	8986	119.10	323.86	5985	59.02	259.36	4162	269.67	6500	353.64

(Contd.)

S.No.	State/UT	1984-85			1985-86			1986-87			1987-88		1988-89	
		Plantation raised	Seedling Distributed	Total expenditure	Plantation raised	Seedling distributed	Total expenditure	Plantation raised	Seedling distributed	Total expenditure	Plantation raised	Total expenditure	Plantation raised	Outlay total
11.	Maharashtra	937	140.42	83.05	245	69.22	39.70	—	—	38.65	989	89.67	2035	160.00
12.	Manipur	1200	15.00	49.34	2325	15.00	78.03	2755	—	92.36	2940	122.65	3050	136.25
13.	Maghalaya	500	10.93	28.96	900	25.63	47.83	1540	24.50	68.80	2386	147.45	5600	90.00
14.	Nagaland	4225	40.00	100.91	5960	32.00	126.32	6700	30.00	139.73	6650	199.90	5820	150.00
15.	Orissa	6846	66.25	199.76	4455	23.38	206.41	9221	15.00	249.03	6533	265.82	6700	300.00
16.	Punjab	2500	100.04	129.56	5000	151.11	237.57	5004	148.00	285.50	5285	280.19	3104	180.00
17.	Rajasthan	8175	102.00	245.24	9000	150.00	200.00	6000	75.00	220.00	5200	263.14	4700	220.00
18.	Sikkim	700	8.00	31.47	1200	10.50	29.61	750	—	27.80	1100	38.18	1380	80.00
19.	Tamil Nadu	4567	132.86	147.41	3590	144.05	119.73	3453	164.54	121.84	4359	132.03	proposal still awaited	
20.	Tripura	3819	20.00	54.42	3012	55.17	74.81	2764	72.84	73.68	2113	101.56	1900	110.00
21.	Uttar Pradesh	7370	408.05	265.71	6106	352.00	325.84	4884	41.63	400.88	6670	369.08	7591	434.83
22.	West Bengal	1810	83.00	39.38	1346	45.37	34.45	—	15.56	47.00	2250	73.95	1691	57.15
23.	Arunachal Pradesh	1052	1.53	17.62	440	0.20	7.70	195	—	6.47	150	2.83	100	2.70
24.	Mizoram	5000	4.00	240.83	6869	308.00	126.43	3797	10.00	110.95	1510	92.28	2500	146.00
25.	Goa	150	—	2.40	210	2.50	5.00	600	—	10.00	400	15.02	500	30.00
26.	Delhi	508	7.65	7.28	261	2.10	8.84	445	2.88	11.63	208	11.25	175	10.00
	Total	93039	2517.73	3429.25	93994	2785.86	3511.42	83627	1494.53	3582.91	80290	3942.18	87433	4154.50

* This component discontinued from October, 1987.

SELECTED BIBLOGRAPHY

BOOKS :

1. Agarwala V.P., Forests in India.
2. Buta Singh, Problems of Prospects of Agriculture in India.
3. Baker, Michal, J. and Tavish Ronald, M.C. Product Policy and Management.
4. Becherman Wilfred, An Introduction to National Income Analysis, Second edition.
5. Benneth, H., Elements of Soil Conservation, Page 44.
6. Chambers, Twentieth Century Distionary, Page 349.
7. Department of Agriculture, Soil Sruvey Manual.
8. Duerr, Teegiwarden, Christiansen, Guttenburge, Forest Resource Management, 1975, Page 10-17
9. Duesenbarry James S. and Preston Lee E., Cases and Problems in Economics.
10. Dwiwedi, A.P. Forestry in India, 1980, Page 21.
11. F.A.O. of the United Nations, Rome (1974), An Introduction to Planning Forestry Development, Page 16-17.
12. F.A.O. of United Nations, Rome (1974), Employment in Forestry, Page 4.
13. F.A.O. of the United Nations, National Forest Policies in Europe.
14. Fernandez, R.R., Dehradun, Gums and Resins.
15. F.R.I. Colleges, Dehradun, Abridged Glossary of Technical Terms.
16. Sir Edmand Hillery, Ecology 2000, The Changing Face of the Earth.
17. Gene W. Grcy, Fredric J. Deneke, Urban Forestry.
18. Gerald Toley, Geoffery Bernarel, Farm and Community Forestry.
19. Gregory, G.R. Forest Resource Economics.
20. Government of India, India, 1980.
21. Harrison, A.J., Economics and Land Use Planning.
22. Dr.(Mrs) Indurkar P.R., Contribution of Forest to the Economy of Nagpur District.

23. Joshi, C.M. and Farasnis, M.S., Forest Labour Co-operative Movement in Maharashtra State.
24. Khoshao, T.N. Environmental Concerns and Strategies.
25. Kaith, D.C. and Erasmus, I. (1958), Wind Erosion, Wind Breaks and Shelter Belts, ICARI, New Delhi.
26. Lal, J.B., Environmental Conservation.
27. Lesli, A.J., F.A.O., Rome, 1971, Economic Problems in Tropical Forestry, Page 15.
28. Lesli, A.J., F.A.O. Rome, Cost Benefit Analysis and Economic Analysis in Forestry.
29. Little, I.M.D. and Mirrlees, J.A., Project Appraisal and Planning for Developing Countries, Oxford and I.B.H. Publishing Co., New Delhi-Bombay-Calcutta.
30. Marshall Alfred, Principles of Economics, Macmillan and Co., Page 434.
31. Mathur, V.P., Forest Management, 1968, Pages 1 and 10.
32. Maharashtra Forests, A Bird's Eye View.
33. National Council of Applied Economic Research, New Delhi, Energy Demand in Greater Bombay.
34. National Council of Applied Economic Research, New Delhi, Domestic Fuels in India.
35. Dr. Negi, S.S., Environmental Conservation.
36. Osmaston, F.C., The Management of Forests, Page 53.
37. Puri, G.S., Meher-Homji, V.M. Gupta, R.K. Puri, Forest Ecology. Phylography of Forest Conservation, Volume I (Second Edition).
38. Dr. Parasnis, S.S., Forestry in Rural Economics with Particular Reference to Agriculture.
39. Pigou, A.C., The Economics of Welfare, Macmillan, London, 1932, Page 10-11.
40. Peirse H. Beresford, F.A.O., 1968, Forests, Food and People.
41. Robers Suagen and Alan Willams, The Principles and Practice of Cost Benefit Analysis.
42. Rangnathan Shankar, Population, The Neglected Factor. Relevant Forestry, Special Issue, "Vaigyanele" June 1981.

43. Serala Devi, Revive our Dying Planet (1982).
44. Sagreiya, K.P., Forest and Forestry, 1967.
45. Sharma, L.C., Development of Forests and Forest Based Industries, 1978, Pages 6, 7.
46. Seth, V.K. and Tiwari, K.P. Forest Labourers Socio-Economic Aspects, A Case Study in Kerala.
47. Sharpex Hendec Allen, 1976, Introduction to Forestry, Page 53, McGraw Hill Inc.
48. Stoddard Charles, H., Essentials of Forestry Practices, 3rd edition.
49. Tiwary, D.N. Forestry in National Development.

REPORTS :

50. All India Dry Zone Afforestation Study Tour, Symposium.
51. Citizens Report, The State of Environment.
52. Energy from Trees—Potential and Problems of a seçond green revolution in India.
53. CSO, Ministry of Planning, Govt. of India, New Delhi, Forest and Logging, Mimeographed Reports.
54. Dutrow George, F., Economic Analysis of Tree Improvement, Astatus Report.
55. F.A.O. of the United Nations, Report on Food and Agricultural Registation, Volume XXXVI, Dec. 1987.
56. F.A.O. of the United Nations, Tropical Forestry Action Plan.
57. F.A.O. of the United Nations, Report of the Council of F.A.O., Ninety fourth session, Rome, 15-26 Nov. 1988.
58. F.A.O. Forestry Paper, Forestry for Local Community Development.
59. F.A.O. of the United Nations, Monitoring and Evaluation of Participatory Projects.
60. Forestry Commission, Bulletin No. 14, Forestry Practice.
61. Forest Division, Ministry of Lands, Natural Resources and Tourism. Tangenia. Trees for Village Forestry.
62. F.D.C.M., Project Report, July 1977.

63. Forest Development Board, Project Report, Technical Report No. 8, May 1973.

64. Government of India, Ministry of Environment and Forests, Indo - U.S. Workshop of Wetlands Mangroves and Biosphere Reserves, 4-7. Jan. 1989.

65. Government of India, Ministry of Environment and Forests, Training Programme for Indian Forest Service Officers in Forest Management.

66. Indian Forest Service Association, Symposium on Role of Forest.

67. Indian Institute of Forest Management, Eco System Approach to Planning and Forest Management.

68. Ministry of Environment of Forests, National Forest Policy, 1988.

69. Ministry of Agriculture, August 1972, Interim Report of the National Commission on Agriculture on Production Forestry, Man Made Forests.

70. Ministry of Agriculture, National Forest Policy, 1952.

71. Ministry of Agriculture, Report of National Commission on Agriculture, 1976, Part IX, Forestry.

72. National Wasteland Development Board, Ministry of Environment and Forest, Government of India, Report of the Working Group for Wasteland Development Sector in the Eighth Five Year Plan.

73. Proceeding of Ist National Symposium Udhagammandation, Sept. 24-26, 1988., Biosphere Reserves.

74. Proceeding of the International Workshop on the Management of National Parks and protected areas in the Hindu Kush Himalaya, 6-11 May, 1985, Kathmandu, People and Protected areas in the Hindu Kush Himalaya.

75. Report of Committee on Fuelwood, National Wasteland Development Board.

76. Sinha, J.N. (1954), Water Sheds or Catchment Barrins of Valley Development Projects, Proceedings of IV World Forestry Congress, Vol. II, Pages 85-89.

Index